DISCLOSURE AND DISCRETION IN ROMAN ASTROLOGY

Disclosure and Discretion in Roman Astrology

Manilius and his Augustan Contemporaries

STEVEN J. GREEN

OXFORD
UNIVERSITY PRESS

OXFORD
UNIVERSITY PRESS

Great Clarendon Street, Oxford, OX2 6DP,
United Kingdom

Oxford University Press is a department of the University of Oxford.
It furthers the University's objective of excellence in research, scholarship,
and education by publishing worldwide. Oxford is a registered trade mark of
Oxford University Press in the UK and in certain other countries

First Edition published in 2014

Published in the United States of America by Oxford University Press
198 Madison Avenue, New York, NY 10016, United States of America

British Library Cataloguing in Publication Data
Data available

Library of Congress Control Number: 2013956985

ISBN 978–0–19–964680–7

To my father, Joe, who will always be a star

Preface

Perhaps surprisingly, the origins of this book can be traced back to an international conference on Ovid (Manchester, 2002), and to the ensuing volume for which I was co-editor along with my Manchester colleagues, Alison Sharrock and Roy Gibson (2002–6). It was during this period that I first started to appreciate more fully the playfulness of the didactic genre, and I was particularly intrigued by the fascinating prospect within *Ars Amatoria* 3 that the speaker addresses two audiences simultaneously while principally promoting the interests of just one. At about the same time, and with my interests growing in the politics of Roman astrology, I started to ponder a series of related questions about the didactic offering of Manilius. Who would be so bold as to purport to teach horoscopic astrology to a general audience at a time when the Emperor was so keen to keep such knowledge and expertise for himself? Why did this poet of politicised didactic escape imperial wrath when his contemporary Ovid had not? Or, is there scope for understanding Manilius' project as a work of significant didactic playfulness against this political backdrop?

I worked up my preliminary thoughts for a paper at the major international conference on Manilius which I co-organised with Katharina Volk (New York, 2008). The conference volume was duly published in 2011. It was a delightful experience to test out my ideas within this stimulating Manilian venture, from which grew the current monograph, and I am very grateful for Katharina's support throughout, even though I appreciate that we hold quite different views of our mutual author.

During the course of writing this book, several readers have offered valuable suggestions to individual chapters, and have helped me out with individual queries. I am very pleased to acknowledge their contribution here: Michael Beardmore, Roy Gibson, Alice König, Daryn Lehoux, David Levene, Jim McKeown, Mirjam Plantinga, John T. Ramsey and Katharina Volk, as well as the anonymous readers for OUP. Special thanks are extended to Robert Maltby for offering comments on a draft of the entire book. The material on astrometeorology formed the basis of a paper delivered at the 'Aratus and the Astronomical Tradition' workshop (St Andrews, 2010), and the sections on Manilius and Vitruvius were given as papers at different research seminars (Trinity Saint David, 2010 and Cambridge and Maynooth, 2013, respectively). I am grateful to audiences at these papers for further valuable suggestions.

I am pleased to acknowledge the support from the Department of Classics at the University of Leeds, which provided two half-year periods of research leave to enable me to work on the project. I would also like to thank Hilary O'Shea and the staff at Oxford University Press for their professional guidance at all stages of the publishing process, and Ben Morales-Correa for giving permission to reproduce his illustration of the Perseus constellation for the front cover.

It goes without saying, of course, that any mistakes remain my own. Moreover, if apology be needed in advance, I am minded to take up a point made by one of the OUP readers that some of the expression used in my discussion of the teacher–student relationship in Manilius strays close to the 'student experience' parlance which is currently fashionable within UK Higher Education managerial circles. This has not been intentional, but it is at times difficult to avoid, and where it does occur, I take some pride in reclaiming this language for ancient didactic poetry.

Finally, this book is dedicated to my close family (Chee, Penny and Joe) and in particular to my late father, Joe, who remains a true star.

S.J.G.

UCL and Singapore
Jan 2014

Contents

1. Introduction 1

PART 1: MANILIUS' *ASTRONOMICA*

2. Manilius' *Astronomica*: A Lesson in Horoscopic Obscurity 11

PART 2: THE RISE OF ROMAN ASTROLOGY AND CAESAR'S COMET

3. The Rise of Astrology in Rome 65
4. The Beginning and End of the Late Republican Astrological Debate: The Politicised Philosophical Posturings of Cicero 75

PART 3: ASTROLOGY FOR THE AUGUSTAN AGE

5. To Have and to Hold: Astrology for an Imperial Age 97
6. Setting Gentlemanly Limits for Imperial Stellar Investigation in Vitruvius' *De Architectura* 109
7. Concession, Abstinence and Abortion: Horace, Virgil, Hyginus and Ovid 119
8. Stars and Storms: The Development of Stellar Causation in Astrometeorology 129
9. Caesar's Comet: The Reinvigoration of a Religious Enquiry 151
10. Turning on the Practitioner: Propertius Meets the Charlatan 173
11. Conclusion: Manilian Dialogues and the Relaxing of Astrological Discretion 189

Bibliography 198
Index Locorum 215
General Index 223

1

Introduction

In light of modern scepticism towards the practice, it is easy to underestimate just how important a role astrology played in the ancient Greek and Roman worlds. The science of astrology may be broadly defined as the 'interpretation and prognostication of events on earth . . . from the measurement and plotting of the movements and relative positions of the heavenly bodies' (Tester 1987: 11), of which there were several distinct branches in antiquity.[1] The most famous and influential was genethliology, or natal/horoscopic astrology, whereby the destiny of an individual (the 'native') was forecast by interpreting the *horoskopos*, the position of the sun, moon and planets in relation to the zodiacal signs at the horizon at the time of birth. The other branch pertaining to the individual was catarchic astrology, whereby the heavenly bodies were consulted in an effort to determine whether a particular date or time was auspicious for conducting a particular activity. Other branches of astrology, by contrast, spoke to wider future phenomena. Meteorological astrology—or astrometeorology—made use of the stars as a means of predicting weather patterns on earth. Moreover, astronomically significant moments such as eclipses or comets were consulted in order to forecast events and conditions which (typically) pertained to entire nations.

Though several branches of astrology are treated in this study, there is a strong focus around horoscopic astrology at Rome. Entering the city from around the 3rd century BC onwards, astrology took hold gradually in Rome, to assume its position as primary mode of divination by the early imperial period. From street diviners peddling their

[1] The categories given here, drawn from both Barton (1994b) and Jones (2007), are not exhaustive. Although we can be confident that distinctions were observed in antiquity, the terminology of categorisation used here comes from modern scholarship.

dubious trade around the Forum and Circus to the hallowed educated experts of the imperial courts, astrologers commanded the attention of plebeian and Emperor alike as potential communicators of truths about their future accessible from a proper reading of the complexities of the stellar world. More specifically, my study centres around a critical period in the development of horoscopic astrology at Rome, broadly speaking from 44 BC to AD 16, and the particular role that Octavian/Augustus played in putting the science at the heart of Roman political life. The ancient evidence points to an intriguingly contradictory if politically comprehensible approach to astrology on the part of Augustus. The Emperor's enthusiasm for securing from astrologers favourable predictions about his own career, and for using astrological symbols to cement his own position of power, is matched by an equally forceful desire to restrict the use of these tactics by others, especially those who might prove to be political rivals. There would seem, then, to be a good deal of truth lying behind Tacitus' neat formulation that astrology 'will always be both forbidden and maintained in our city' (*in civitate nostra et vetabitur semper et retinebitur*, *Hist.* 1.22). My study is the first to take seriously this imperial complex as a key to understanding the diverse ways in which contemporary commentators handle the volatile topic of astrology in their writings. It aims to show how Roman writers engage in elaborate discourses of disclosure and discretion as they find creative ways to celebrate the power of astrology while bringing their discussions within the parameters of politically acceptable stellar enquiry in the early empire. Put another way, my study stresses the importance of a particular conceptual framework for a fuller appreciation of the varied negotiations on astrology during this key early imperial period, with particular emphasis on the understudied astrological poem of Manilius, which provides the clearest window onto this phenomenon.

Because of the importance of Manilius to my argument, I make no apology for resisting a strictly chronological approach in this study and starting with an investigation into the late Augustan poem, *Astronomica* (Chapter 2): this is, I maintain, the work which provides the clearest signals of the tension between astrological knowledge and political power under the early empire, and of the author's own negotiation of this tension. Manilius' didactic project purports to teach the lay reader how to construct a horoscope. Though scholars have long recognised the inadequacies of Manilius' poem as instruction in this regard, I take this much further by arguing for a systematically deliberate strategy on the part of the poet. By building on a short contribution elsewhere,[2] and by paying particular attention to the deterioration in the relationship between didactic teacher and student in the poem, I argue for the legitimacy of reading the *Astronomica* as a deliberately failing lesson. The reason for this poetic strategy relates directly to the contemporary nervousness that surrounds the issue of practising astrology in the early empire. Rather than being the first Roman work on astrology, therefore, Manilius' poem is central to an appreciation of a discourse of disclosure and discretion on the topic which develops over the course of the Augustan reign.

With the most elaborate exponent of astrological discretion so revealed, I then take a broadly chronological approach by charting the beginning and progress of this discourse in the age of Octavian/Augustus. I start by offering a brief history of the rise of astrology in Rome up until the appearance of Caesar's comet in 44 BC, an event which appears to have marked a turning point in astrology's (imperial) credibility (Chapter 3). Analysis of astrological passages across two contemporary works of Cicero, *De Divinatione* and *De Fato*, reveals a softening of stance towards the practice which, I argue, may reflect Cicero's recognition of the fast-changing political status for astrology (Chapter 4).

This sets the scene for a variety of creative, politically oriented approaches to astrology in writers under Augustus, the focus of Chapter 5. After a brief overview of Augustus' double standards in his use of horoscopic astrology—the Emperor's simultaneous exploitation of astrology for his own ends and restriction of its use by others (Chapter 5)—I consider a series of Roman writers who find their own ways into this imperial complex of astrological celebration and discretion. Vitruvius provides the earliest example of astrological negotiation under Augustus when he chooses to celebrate the stellar world and the science of astrology while conceding its practice strictly to selected hallowed experts (Chapter 6). This pro active discretionary stance towards disclosure of astrological knowledge proves popular with authors such as Horace, Virgil, Hyginus and Ovid, who either offer direct warnings to others against the consultation of astrologers or actively showcase their avoidance of the topic altogether (Chapter 7). Other writers, perhaps more creatively, identify the topic of astrometeorology as a safe and appropriate means of promoting the predictive power of the stars

[2] See Green (2011).

without incurring official disfavour (Chapter 8). Another safe topic around which authors felt comfortable to focus praise of astrology was the (dynamic) issue of the catasterism of Julius Caesar (Chapter 9). Taking a rather different stance, Propertius draws a distinction between the science (worthy of respect) and the practitioner (who may be fraudulent) and, through the figure of Horos, finds a safe and amusing means of advertising the power of astrology while poking fun at one of its less credible exponents (Chapter 10). Finally, the conclusion (Chapter 11) stands back and takes stock of the uniqueness of the age of Augustus as a period of creative literary engagement with astrological politics in a way that avoided outright criticism of the science.

The 'disclosure' of the title, then, relates to the various ways in which the power and influence of astrology is advertised in contemporary writers, both by expressions of its efficacy and by exploration into branches of the science which are felt not to be politically sensitive. The modes of 'discretion' operated by these same authors towards horoscopic astrology, however, take many forms: direct expressions of the exclusivity of the art, active (and sometimes teasing) avoidance of the topic, due attention to the avoidance of inappropriate displays of horoscopic knowledge and, most elaborately in the case of Manilius, the practical handbook that perpetually frustrates.[3]

In focusing on the complex place of astrology under Augustus, my study aims to make a distinctive contribution to existing scholarship focussed on the relationship between power and knowledge in the early empire, at a time when control of knowledge was passing from late Republican social leaders into the hands of academic experts working in a potentially reciprocal relationship with the Emperor.[4] In general terms, my study follows in the footsteps of König and Whitmarsh (2007) in analysing the ways in which specific knowledge-bearing texts play host to the complex relationship between knowledge

[3] One might raise a methodological objection here that it was astrological *practice*, rather than astrological *knowledge* per se, which appears to have been the focus of imperial unease. However, as Ripat (2011: 121–8) has successfully shown, the line between knowledge-bearers and knowledge-users in Roman astrology, and indeed between practitioners who considered themselves as 'astrologers' and those that did not, was a very thin one. Within this context, I maintain that, even outside the overtly didactic confines of Manilius' poem, a contemporary author would need to have been mindful of the fact that any competent publicised discussion of horoscopic astrology ran the risk of setting imperial nerves on edge.

[4] See especially Wallace-Hadrill (1997). Moatti (1997) likewise attributes the growth in categorising technical knowledge in the early empire to the collapse of late Republican politics and society.

acquisition and imperial power 'to reveal beneath their dispassionate surfaces something of what it is possible to say or to think within the societies and disciplines from which they arise' and to say something about the social and political self-positioning of the author within this relationship.[5]

But this study owes a significant debt to two particular studies on Roman astrology. The first is Tamsyn Barton's *Power and Knowledge* (1994a), which is the first work to give serious attention to the complex interplay between astrology and imperial power, and the ways in which powerful individuals worked to create consensus on its truth.[6] Focusing on the astrological treatises and self-dubbed 'handbooks' from the Second Sophistic (*c.* late 1st to mid 3rd centuries), Barton detects a range of strategies by which authors complicate their astrological instruction and create frustration in the mind of the reader/student: calculations are asserted without clear rationale; incompatible systems of calculation and/or various means of calculation are presented uncritically; information is overloaded; there is general improvisation and irresistibility to being pinned down; meaning is often deferred; and complication seems to cancel out advice previously given.[7] Barton argues that these difficulties are a deliberate strategy on the part of the authors to create confusion, and she offers overlapping explanations to account for this. Astrology may be deliberately constructed as a hierarchical game between an all-knowing master and a flagging pupil, who is meant to feel inadequate and recognise the seemingly unbridgeable gulf between himself and the instructor. Or perhaps this is a window onto the competitive, financially driven world of the astrologer, who must maintain an air of complication in order to create a sense of expertise, and must retain enough flexibility in his astrological system to pander to the needs of a diverse and paying clientele.[8]

Barton, then, concentrates on the mid-to-late imperial writers of astrological treatises as self-constructed figures of power in their

[5] König and Whitmarsh (2007) 7.

[6] While Frederick Cramer's *Astrology in Roman Law and Politics* (1954) may lay claim to being the first serious treatment of the interaction between astrology and politics in Rome, and while it continues to be an indispensible aid to any such study, there is a tendency to treat sources at face value without giving significant attention to the motivations of their authors and the intended audiences.

[7] See Barton (1994a) 71–94 ('Astrology in Practice').

[8] See Barton (1994a) 82–94.

own right.[9] By contrast, I am focusing on an earlier time period and on writers—not primarily astrological practitioners, as far as we can tell—who negotiate levels of disclosure on astrological knowledge and position themselves in respect to the Emperor, who is tacitly acknowledged as the (sole) locus of power. In so doing, I must acknowledge a second debt, this time to Geraldine Herbert-Brown. In a fascinating piece centred on the stellar material in Ovid's *Fasti*, Herbert-Brown sketches out the effects on contemporary authors and works of Augustus' contradictory approach towards astrology during his age.[10] Whenever I return to Herbert-Brown's piece, I am struck by the extent to which the current monograph seems to have generated from the seeds of investigation contained within that piece, albeit with some important qualifications.

First, Herbert-Brown treats Augustus' astrological legislation of AD 11 as a decisive turning point in the Emperor's level of tolerance towards astrological disclosure, whereas I maintain that the edict of AD 11 merely lent legislative muscle to an imperial approach towards astrology that can be seen from at least the time of the expulsion of astrologers from Rome in 33 BC.[11]

Secondly and more importantly, Herbert-Brown argues that the validity of astrology continued to be questioned within the Augustan age, providing as evidence the diatribe of Ovid's teacher, Arellius Fuscus, recorded in Seneca's *Suasoriae* (4.1–3).[12] It may indeed be true that debates about astrology remained part of the intellectual framework during this time, but I would draw attention to the fact that such debates are only *recorded* at a time outside the Augustan reign.[13] What must be observed is the remarkable fact that, despite the various ways in which astrology is handled during the Augustan period, there is no (extant) written criticism of astrology—the harsh verdict of 'Cicero' in *De Divinatione* 2.87–99 (44 BC), who apparently felt at ease in committing his

[9] In so doing, she draws on Michel Foucault's definition of power as a complex network of relationships which are being acted out and negotiated at every level of society (here, between self-appointed expert and flagging student/reader); see Barton (1994a) 20–5, and more generally Dirks, Eley and Ortner (1994) 4–5, 6–11.

[10] See Herbert-Brown (2002) 113–21; this offers a more nuanced treatment of this period than Barton (1994a: 47–54), whose focus is, as we have said, on a later time period.

[11] See Chapter 5.

[12] See Herbert-Brown (2002) 120.

[13] Moreover, there is an argument to suggest that at least some of Seneca's 'recollection' of Fuscus' diatribe is coloured by his own contemporary, Tiberian setting; see Chapter 11.

views to writing, is nowhere replicated under Rome's first Emperor. For overt criticism, we have to wait until the time of Tiberius/Gaius (Sen. *Suas.* 4.1–3) and the flamboyant figure of the sceptic Favorinus (AD 85–165, ap. Gel. 14.1).[14] This is quite remarkable, in view of the fact that imperial sponsorship of particular religious practices did not, of course, guarantee compliance from state-sponsored contemporary authors, who often found ways to re-establish debate and reduce the Emperor to just another voice in the crowd. For example, Augustus seized upon both animal sacrifice and augury as key symbols of his identity as an august, pious reviver of Roman religious institution, and yet contemporary authors continue to keep open the spirit of debate by, respectively, raising the spectre of Pythagoras and questioning the benign mythological underpinning to man's relationship with birds.[15] If one accepts the readings in the present study, then, horoscopic astrology would appear to provide a novel example of compliance, even from the perennially mischievous Ovid, whereby a creative space was explored within commonly respected imperial boundary lines: this is a game of disclosure, but one which is mixed appropriately with due omission, complication and silence—indeed, ostensible displays of silence.

With the scene now set, there is no more fitting place to start our enquiry than with a self-purported lesson on our very topic, offered by Manilius.

[14] See further Chapter 11.

[15] See Green (2008) and (2009).

Part 1

Manilius' *Astronomica*

2

Manilius' *Astronomica*

A Lesson in Horoscopic Obscurity

Manilius is the author of a five-book poem about astrology entitled *Astronomica*. About the life and personal circumstances of the author we know almost nothing, and attempts to construct a biography for him have been necessarily speculative.[1] We are slightly better placed to determine the date for *Astronomica*, as the poem provides a *terminus post quem* of AD 9 with its unambiguous reference to the Battle of Teutoburg Forest (1.898–903). Beyond this, however, debate continues as to whether the poem is entirely Augustan (i.e. written before AD 14), entirely Tiberian (i.e. started after AD 14), or a work which straddles the reign of both Emperors. Not unrelated to this is the issue of whether the poem was completed before or after the Augustan legislation of AD 11 which was designed to curtail certain forms of astrological disclosure.[2]

[1] Pliny (*Nat.* 35.199) mentions a 1st-century BC astrological writer named Manilius Antiochus, a reference which may invite connection to our author. For various attempts to construct a biography and sphere of influence for our author, see, for example, Scarcia in Feraboli, Flores and Scarcia (1996–2001): 1.xix–xxiii; Goold (1961) 171–2; Volk (2009) 1–6.

[2] It may be fair to say that there has been a very general consensus in recent times towards a dating for the work of *c.* AD 9–16, but this has been challenged by Volk, who also provides an excellent history of the debate; see esp. Volk (2009) 137–61; Volk (2011a) 4–5. As will become clear, my own reading depends essentially on an *imperial* context for the poem and, as such, does not require a decision as to the specific Emperor addressed at any one point. The debate over the poem's relationship with the Augustan legislation of AD 11 need not detain us either. The poem was produced either in the wake of the legislation of AD 11 or, at the very least, in the context of an imperial nervousness of unsolicited astrological disclosure that goes back to the beginning of the Emperor's reign; for which, see Chapter 5.

In terms of the poem's genre, *Astronomica* is unanimously classified by modern scholars as a didactic poem. But what does this mean? First, it is important to clarify that literature that we readily label 'didactic' was often grouped in antiquity, by metre, with epic. The first evidence for didactic as a distinct genre of poetry dates from late antiquity (Diomedes the grammarian, 4th–5th centuries AD), and it is a matter of scholarly debate as to whether poets of the late Republic and early empire were consciously operating within such a generic category.[3] Even if not treated as a specific genre, however, it is generally agreed that there are sufficient defining qualities to such literature to set it apart, at least, from 'conventional' narrative epic. Scholars have proposed various slightly differing sets of defining features for didactic poetry, but all agree that there must be an internal dialogue between a single instructor and an individual student/set of individual students (specified or otherwise); a teacher with a clearly purported intent to offer systematic instruction in an area of knowledge (be it practical, scientific or moral/philosophical) which is capable of being taught.[4] Assessed against these central criteria, Manilius' poem is most definitely didactic, even if its didactic voice takes a little while to announce itself.

What follows is a detailed discussion of Manilius' poem, focussed around the relationship established between teacher and student and the development of that relationship as the poem progresses. Though this relationship has been the subject of previous scholarly discussions, it is typically the teacher-poet and his intentions which have

[3] See Toohey (1996) 5–7 and esp. Volk (2002) 26–34. Manilius' own select lists of literary predecessors (2.1–48, 3.5–26) offer no real clues as to any perceived generic affiliation: there are allusions to poems that we might label didactic—such as Hesiod's *Works and Days* (2.19–24) and the pharmaceutical works of Nicander (2.44–5)—but the purpose of these lists is thematic rather than generic, as Manilius seeks to contrast his own novel enterprise with the well-trodden paths of others (2.49–59, 3.26–35).

[4] See Toohey (1996) 2–5; Fowler (2000); Volk (2002) 34–43. The word 'intent' is important here. With regard to the didactic poet Aratus, for example, Cicero claims (*De Orat.* 1.69, *Rep.* 1.22) that his expertise is in poetry rather than astronomy. But the relative expertise of the author in his field of instruction has no direct bearing on the author's intention/willingness to engage in the pedagogical process of teaching. Besides, if authors such as Manilius do not specifically state their expertise in their field of study, they are following in a tradition of didactic poets who stake authority in terms of poetic/divine inspiration rather than knowledge/experience of subject matter; Sharrock (2005) 245–9.

been prioritised in such analyses.[5] In contrast, I want to take a much more student-focussed approach to the text, from the moment he/she is identified early in Book 1 until the end of the poem/lesson in Book 5. I also adopt a more strictly linear approach to the text than previous scholars, in the hope that this may allow us to assess more faithfully the structure of the teacher's lesson, and to detect more clearly the development in his relationship with his addressee as the poem/lesson progresses. And it will become clear that I am arguing for a *deterioration* in the relationship between student and teacher during the poem, with Book 4 acting as a pivotal moment.

ESTABLISHING A POSITIVE DIDACTIC FRAMEWORK: SPEAKER AND ADDRESSEE(S) IN BOOK 1

The opening lines to the poem duly set out its thematic remit (1.1–6):[6]

> carmine divinas artes et conscia fati
> sidera diversos hominum variantia casus,
> caelestis rationis opus, deducere mundo
> aggredior primusque novis Helicona movere
> cantibus et viridi nutantis vertice silvas
> hospita sacra ferens nulli memorata priorum.

> By song/ incantation I go forth to draw down from heaven divine skills and the stars, confidants of fate which lend variety to the diverse fortunes of men—the work of celestial reason/ calculation—and to be the first to stir with new strains Mount Helicon and the woods nodding with their leafy tops, bringing strange sacred gifts untold by any of my ancestors.

These opening lines are particularly dense in terms of intertextual resonance and magical, scientific and Gigantomachic overtones.[7] What concerns us here, however, is Manilius' stated purpose and motivation:

[5] The most valuable study to date is Neuburg (1993) 257–82, largely followed by Volk (2002) 198–209. While both recognise problems in Manilius' pedagogical techniques, their overriding impression is one of a teacher who is both helpful and encouraging to his student.

[6] The text of Manilius is taken from Goold (1985). All translations are my own.

[7] See e.g. Wilson (1986).

he aims to talk about the stars in their horoscopic role as well as the art of astrology itself, and he is motivated primarily by the poetic novelty of the enterprise. He does not specify at this stage an intention to teach astrology to any specific individual: he may be 'drawing it down' from heaven, but no hint has yet been given as to a target audience.

The next lines bring in a specific addressee (1.7–10):

> hunc mihi tu, Caesar, patriae princepsque paterque,
> qui regis augustis parentem legibus orbem
> concessumque patri mundum deus ipse mereris,
> das animum viresque facis ad tanta canenda.

> You, Caesar, first citizen and father of the country, who rule a world obedient to your august laws and, as a god yourself, deserve the heaven that was granted to your father—you give me heart for this undertaking and create in me the strength to sing such lofty themes.

We need not detain ourselves here with the indeterminable question as to whether *Caesar* refers to Augustus or Tiberius, or indeed whether the *Caesar* of later books is the same as this one.[8] Manilius makes it clear that an Emperor is the poet's inspiration for his undertaking, and that he is a most fitting dedicatee: the poet is bringing down skills from the *mundus* (1.3), and the Emperor is fully deserving of this gift in the light of the fact that the *mundus* is already a family concern (1.9). Manilius will flatter the Emperor further by suggesting implicitly that he has provided the peaceful conditions essential for such an undertaking (1.11–13). So far, then, Manilius has fashioned his poem as an exploratory exposition on the astrological role of the stars, for which the Emperor provides a fitting dedicatee in terms of his stature, divine trajectory and role as guarantor of peace. This does not as yet meet our central criteria for a didactic poem.

After an anthropology of the invention of astrology (1.25–112), Manilius starts his stellar instruction with a description of the universe and the various theories relating to its origins (1.118ff.), before an addressee is specifically evoked again (1.194–5):

> nec vero admiranda **tibi** natura videri
> pendentis terrae debet.

[8] In this instance, the use of *augustis* (1.8) points most naturally to Augustus, though the reference to *Caesar* later at 4.935 might point to a change in Emperor that has occurred between the times of composition for these two books; for divergent views on the identity of these Caesars, see esp. Neuburg (1993) 244–57; Volk (2002) 200–1; Volk (2009) 137–61.

> Nor indeed should the principle of the earth's suspension be seen as a source of wonder **to you**.

To whom does *tibi* refer? It cannot realistically refer back to the *Caesar* of 1.7, as it would be quite out of turn with the reverential tone of 1.7–10 for Manilius to suggest that a great ruler, destined for the heavens and equated with a god himself, might be dumbfounded by the concept of the earth's suspension. Nor does the directive nature of 1.194–5—note *admiranda* and *debet*—fit an address for the Emperor to whom a more deferential tone had been adopted in 1.7–10. Rather, albeit unexpectedly, *tibi* draws attention to a different type of addressee, an unspecified individual whose naïve bewilderment towards the cosmos renders him in need of instruction from the self-appointed teacher-poet. Here, then, is our first glimpse of the target student for the work.[9]

It is important at this juncture to consider the relationship between internally constructed student and external, contemporary reader—or, to use terminology proposed by Sharrock, between 'Reader' and 'reader', respectively. Speaking about Ovid's *Ars Amatoria*, Sharrock suggests that the unspecified nature of the addressee in *Ars* (any Roman: *Ars* 1.1 *si quis in hoc . . . populo*) facilitates slippage between internal student (Reader) and external reader (reader), although the reader still has a certain amount of choice as to when and to what extent he wishes to identify with or disassociate himself from the Reader.[10]

Sharrock's approach is fruitful for our study of Manilius, as the nature of Manilius' student is even less well defined than it is in Ovid's *Ars Amatoria*: he is strictly only a (Latin-speaking) individual living in the northern hemisphere (1.242, 373–81). I would argue that the slippage between Reader and reader is even more tempting, if not ineluctable, in Manilius. As will become clear in the following chapters, for the contemporary reader of Manilius astrology was both a very popular science and one whose technical underpinning was obscure to all but a handful of apparent experts. It is, of course, very difficult to ascertain what sort of topics an ancient Latin-speaking/Roman reader would have found difficult to comprehend, and any judgement is inevitably influenced by one's own contemporary sense of difficulty.

[9] The delayed articulation of the teacher–student relationship in Manilius' didactic poem should not in itself concern us: note especially that Aratus only begins to address his student at Arat. 75.

[10] See Sharrock (1994a) 7–8, 11–17. The addressee of Aratus' *Phaenomena* is left similarly anonymous; see Bing (1993) 99.

In the light of the popularity of astronomical texts such as Aratus, I am certainly prepared to accept that the average elite Roman reader might have been more conversant in the names, general shape and relative positioning of constellations in the sky than the average modern western reader.[11] For this reason, Manilius' first book, which is heavily indebted to Aratus and deals with the positions of constellations in both hemispheres, might have been a particularly easy-going read for the ancient reader.[12] But the complex variables and mathematical calculations required for casting a horoscope, an essentially foreign art whose terminology regularly resists translation into Latin (1.20–4, 3.31–42, 4.431–42), would, I suggest, have been a challenge to all but the small group of acknowledged experts—hence, from Book 2 onwards, Manilius appeals to the heavily cerebral nature of the task ahead.[13] I would argue then, from at least Book 2 onwards, for a strong convergence between Reader and reader. Moreover, for any reader with (mild) subversive tendencies, taking on the role of Manilius' student might have proven particularly attractive in the light of the imperial attempts to restrict dissemination of astrological knowhow beyond the ruling class. The very process of reading a self-purported manual on astrology can be seen, therefore, as a subversive political act, as one colludes with the teacher to find out the means to practise something which is prohibited and, by such means, to gain access to a key source of (exclusive) imperial power.[14] To my mind, then, the pool of potential contemporary Roman readers who might not want to associate themselves with the student constructed in the text is very small: only the experts of astrology, and those too scared to read the book behind closed doors, are likely to have excluded themselves.

From these lines, then, emerge the familiar pairing of the didactic teacher and his student, and it is this relationship which appears to govern the direct addresses from this point in the poem.[15] Some comments

[11] For the view that Aratus may have achieved the status of an established school text designed to provide a gentle introduction to the stars, see Lewis (1992) 113–17.

[12] For the strong influence of Aratus in Manilius' first book, see Abry (2007a).

[13] See below, n. 28.

[14] For the politics of early imperial didactic in general, see Sharrock (1994b), esp. 103–8.

[15] I leave out of the following discussion of imperatival expressions Manilius' use of the first person plural—e.g. in Book 1, 209 *aspicimus*, 216 *conspicimus*, 239 *sub pedibusque . . . nostris*, 242–5 *occasus nostros . . . nos . . . vocamus*—as it is impossible to tell whether he is envisaging the Emperor and himself, the didactic student and himself, all three individuals, or (Ro)mankind in general. I also rule out of the enquiry the sort of poetic apostrophising that is designed simply to lend variety to a list; cf. e.g. 3.568, 571, 576, 4.336, 360.

might equally be made to both addressees simultaneously, though for different reasons: for example, Manilius' intention to relate the fixed constellations of the heavens (1.255–6 *nunc tibi signorum lucentis undique flammas/ordinibus certis referam*) might be designed both to bring pleasure to the Emperor and to offer instruction to the ignorant student. But consider the teacher's words at 1.373–4:

> **aspice nunc** infra solis surgentia cursus
> quae super exustas labuntur sidera terras.
>
> **Look now** at the constellations rising below the sun's course, which set above parched lands.

Manilius here either exhorts or instructs his addressee to look directly for the southern constellations.[16] To which of our two addressees is this direct appeal more appropriate? An answer to this question might be better informed by first considering a study of levels of address in didactic poetry conducted by Gibson (1997) and Sharrock (1997). Gibson takes a sample of imperatival expressions from selected books of Latin didactic verse and prose.[17] The statistics generated allow him, with due caution noted, to draw some very interesting conclusions. First, prose texts generally favour passive/impersonal forms of address and/or avoid directly addressing a reader altogether: second-person directives are rare. Gibson adduces that the type of imperatival form adopted might be directly related to the intended elite audience for the work. For example, the farming manuals of Cato, Varro and Columella relate to a practice for which a rich landholder would traditionally employ slaves to work the land. For this reason, it would be entirely appropriate for the writer to express instruction passively—'let the fields be tilled' rather than 'till the fields'—and to downplay the role of the addressee in the execution of the instruction; in this way, the poet establishes a discrete relationship between himself and his addressee, rather than a position of power over him. Poetic didactic texts, by contrast, favour active forms which address the reader directly, where a clear sense of hierarchy between speaker

[16] Most of the southern constellations would have been at least partly visible to those from the Northern hemisphere; see Volk (2009) 37–9. For more implicit appeals to the student to observe the skies, cf. 1.260 *omnia* [sc. *sidera*] *quae possis caelo numerare sereno*, 1.552 *hinc igitur quodcumque supra te suspicis ipse.*

[17] From poetry: Lucretius, Virgil's *Georgics*, Ovid's *Ars Amatoria*, Grattius and Columella 10; from prose, Cato, Varro, Columella, Celsus and Scribonius Largus. Sharrock (1997) 110–11 widens the field by including also Vitruvius. Neither scholar treats Manilius in any detail (see later).

and addressee is established, most notably in Ovid's *Ars Amatoria*, where the speaker masquerades as the boastful erotic expert over young hopefuls.[18]

The interpretations that one can draw from such statistical data are, of course, fraught with complication.[19] Nevertheless, Gibson's study is illuminating, especially if one thinks less in terms of a distinction between prose and poetry and more in terms of the activity for which instruction is given, and whether or not it is an activity directly carried out by the elite. It is remarkable that both the prose farming manuals and the poetic farming treatise of Virgil's *Georgics* make scant use of direct imperatival expressions.[20] This might suggest that it is the servile nature of Roman farming, rather than the literary format of the instruction, that is the most important factor in the writer's choice of appropriate imperatival expression.

This is a fruitful note on which to return to the imperative *aspice* at Man. 1.373. As is explained more fully in the next section, the Emperor, be he Augustus or Tiberius, was not actively engaged in the consultation of the stars in the pursuit of astrological readings: both had the resources to consult their own select experts in the science.[21] Neither the directness of the tone nor its sentiment, then, suits an address to the Emperor. Rather, Manilius' instruction speaks to the type of reader outside the imperial circle who does not have access to reliable experts in the field. Moreover, as an additional allurement to the budding student and in line with Manilius' lofty claims for innovation, our author espouses the highest and purest form of astrological learning, one which demands consultation of the stars themselves, rather than a reliance on second-hand, commercially available stellar charts and almanacs.[22] Gibson's study, then, might be used to reinforce the impression that

[18] See esp. Durling (1958).

[19] See the response from Sharrock (1997).

[20] Although even these rare direct imperatival expressions, it has been argued, may operate as part of a conceit that the master works the fields through the means of servile prosthesis; see Reay (2005) with regard to Cato.

[21] For Octavian's consultation of Theogenes early in his career, see Chapter 3. For the infamous Thrasyllus, the first Roman imperial-court astrologer under Tiberius, see Chapter 5. For Augustus' conscious reassignment of technical knowledge away from social leaders towards academic experts, see Wallace-Hadrill (1997).

[22] For the real world of the practising astrologer, see Jones (2007) 309–11 and Chapter 10.

Manilius' instruction is directed principally to a wider studentship, with the speaker himself firmly established in the position of expert instructor.

The sense of an expert guiding an enthusiastic but naïve student continues (1.458–60):

> tu modo corporeis similis **ne quaere** figuras,
> omnia ut aequali fulgentia membra colore
> deficiat nihil aut vacuum qua lumine cesset.

> Only you, **do not look for** shapes like bodily shapes, (thinking)
> that all their parts shine with equal brilliance, that nothing is
> deficient or missing through absence of some illumination.

This amounts to the poet's first instance of didactic 'retuning'. Manilius has previously described the anthropomorphised constellations in the Northern and Southern hemispheres respectively (1.275–455). In his role as astute teacher, however, he realises that this may lead to the false impression that all familiar aspects of the relevant character or animal will be discernible in the sky, an impression which he must now rectify. Such retuning of the didactic message arguably serves a double function. Most obviously, it seeks to demonstrate the teacher's sensitivity to the level of knowledge in his student and a concern for the student's successful progress. By the same token, it increases the student's attentiveness towards the teacher, as the student can never be sure that what he has just learnt is as straightforward as it first appears.[23]

The teacher courts the attention of his student in similar fashion a little later (1.474–7):

> et, **quo** clara **magis possis cognoscere** signa,
> non varios obitus norunt variosque recursus,
> certa sed in proprias oriuntur singula luces
> natalesque suos occasumque ordine servant.

> Furthermore, (I offer this) **so that you might better recognise**
> the bright constellations: they know no variation in their settings
> or their returnings, but each one rises in fixed fashion with its
> customary lights, and they keep in order their risings and settings.

[23] For use of this didactic technique in Ovidian didactic, see Green (2004) 52–3 on *Fast.* 1.49–52. For more 'sinister' use of this technique later in the poem, see the discussion on Book 4.

Here in 1.474 the teacher makes explicit his kindly pedagogical disposition,[24] and his determination to ram home the reassuring message that the constellations rise and fall in fixed fashion each year is expressed in different ways across three lines (1.475–7): a notion of fixed patterning implies, after all, that only one set of rules need be learnt.

As Book 1 moves towards its conclusion, the direct instructions gather relative pace. There is another exhortation to learn the science behind the stars in order to avoid naïve feelings of amazement (1.557–60):

> **ne mirere** vagos partus eadem esse per astra
> et mixtum ingenti generis discrimine fatum,
> singula cum tantum teneant tantoquo ferantur
> tempore.
>
> **Do not be amazed** that offspring with divergent fortunes can emerge from the same stars, and that destinies vary with huge differences of kind, given how large is the space that each constellation occupies and over how much time each constellation revolves.[25]

Other direct exhortations also bear witness to the apparent enthusiasm of the student, to whose desire for knowledge the teacher purports to speak (1.648–9):

> alterius fines **si vis cognoscere** gyri,
> circumfer facilis oculos vultumque per orbem.
>
> **If you wish to recognise** the boundaries of the other circle (i.e. the horizon), let your eyes and gaze freely rotate in a circle about yourself.

In summary, Book 1 clearly develops an internal didactic relationship between an apparently well-informed teacher and an apparently

[24] For various explicit comments from the didactic teacher relating to his kindly pedagogical disposition, cf. e.g. Ov. *Ars* 3.443, *Fast.* 1.58 *ne fallare cave*, 2.151. Such kindly disposition towards the student also manifests itself in Manilius' sense of duty (often using a gerundive) in providing the full picture of knowledge; cf. 1.120 *ipsa mihi primum naturae forma canenda est*, 1.560A *restat ut aetherios fines tibi reddere coner*, 1.811 *implenda est mundi facies*. For later concerns from Manilius to deliver the student from error, cf. 3.389, 506: though in these cases, as I argue later, the sentiment has a hollow ring to it.

[25] For the favourite injunction *ne mirere* in Manilius, cf. 1.859, 904, 2.423, 951; cf. also 2.534, where he applies it to himself (*quid mirer . . . ?*). See further Neuburg (1993) 261–3.

enthusiastic but less-knowledgeable learner—whom I am prepared to identify quite strongly with the external contemporary Roman reader, at least from Book 2 onwards. The teacher appears helpful and sensitive to the needs of the student. Nor are the instructions given by the teacher difficult for the student to perform, as they often involve little more than observing the sky and applying basic reasoning to explain what he sees.[26] But by the same token, the teacher has not strictly taught anything yet about *astrology*. All Manilius has done so far is to map out some of the visually discernible features of the cosmic backdrop for a study of astrology. The student has certainly not yet been introduced to the sort of cosmic variables and mathematical calculations necessary in order to cast a horoscope, a lesson reserved for the subsequent books.

INCLUSION OR EXCLUSION? COMPLICATIONS TO THE LESSON IN BOOK 2

After a respectful survey of literary predecessors who have handled familiar themes (2.1–48), Manilius once again offers self-congratulation for the novelty of his astrological enterprise (2.49–59) before setting out a discernibly Stoic vision of the workings of the cosmos: the universe, under the control of an omniscient deity, is a sentient and rationally organised 'body' (2.60–81); in terms of astrology, the god has created a direct and constant connection between particular constellations and the lives of human beings born under them (2.82–6),[27] and ready proof of a connection between stars and mortal life can be found in the wide-ranging influence of the moon (2.87–114).

So far, this may be regarded as an elaboration of the world-view that the poet projected more briefly in Book 1 (2.47–54). But at this point, Manilius extends the philosophical discussion into new territory which has a sudden and profound effect upon the teacher–student relationship (2.105–8):

[26] For example, when Manilius implicitly instructs his student to observe that the constellations are not all visible at the same time (1.215–17), he is inviting him to apply the recent knowledge about the spherical nature of the world (1.194–214); see further Neuburg (1993) 266, esp. n. 44.

[27] Manilius initially suggests that the stars affect all animate life on earth (2.83 *terrena animalia*), before narrowing the focus specifically to human beings (2.86 *gentibus*).

quis dubitet post haec hominem coniungere caelo,
<cui, cupiens terras ad sidera surgere, munus>
eximium natura dedit linguamque capaxque
ingenium volucremque animum, quem denique in unum
descendit deus atque habitat seque ipse requirit?

Who after this can doubt that heaven is joined with man, <to whom, in his desire for earth to rise to the stars>, nature has given an outstanding <gift>—speech, keen intellect and a winged mind—and into whom alone a god ultimately descends and resides, the god himself seeking himself?

Here the Stoic conceit of a universe that is sentient and whole has implications for the student of astrology, and happy implications at that: everyone, just by virtue of being human, it would seem, is endowed with divinely given intellect that can help him comprehend the cosmos. It is certainly fitting at this stage for Manilius to turn to the topic of intellect: whereas the instruction in Book 1 had chiefly concerned autopsy, the more abstract nature of the information presented from Book 2 onwards is going to require the use of the mind rather than the eyes. Put another way, the visual, Aratean frame of reference that might have guided the student in Book 1 must now be put aside in favour of a more conceptual approach to the stars, as Manilius' philosophical didactic project starts to take on the same cerebral dimensions as Lucretius' *De Rerum Natura*.[28]

The sentiment is reinforced a little later with embellishment, where the deity and nature appear as interchangeable agents of benevolence (2.115–25):[29]

quis caelum posset nisi caeli munere nosse,
et reperire deum, nisi qui pars ipse deorum est?
quisve hanc convexi molem sine fine patentis
signorumque choros ac mundi flammea tecta,
aeternum et stellis adversus sidera bellum

[28] For the overriding motif of autopsy and the visibility of the object of study in Aratus, evident in the poem's title (*Phaenomena*), see e.g. Hunter (1995); for appeals to autopsy in Manilius Book 1, see n. 16. For specific appeals to (acute) mental reasoning in Book 2, cf. 2.157–8 *sagaci . . . animo*, 203–4, 234–5, 693, 788 *animum compone sagacem*, 842 *memori . . . mente*, 898–9 *sub corde sagaci/conde*, 927; for specific advice to prioritise mental over visual reasoning in Book 2, cf. 2.203–5, 297–341; for such appeals in Book 3, see n. 39. The use of *sagax animum* sounds a particularly Lucretian note; cf. Lucr. 1.50, 2.840, 4.912; cf. also Lucr. 1.130 *ratione sagaci*, 5.420 *sagaci mente*.

[29] For the interchangeability of God and Nature as universal creators in Manilius, cf. e.g. 2.435, 723 (*natura*); 2.475, 768 (*deus*).

> [ac terras caeloque fretum subiectaque utrisque]
> cernere et angusto sub pectore claudere posset,
> ni sanctos animis oculos natura dedisset
> cognatamque sibi mentem vertisset ad ipsam
> et tantum dictasset opus, caeloque veniret
> quod vocat in caelum sacra ad commercia rerum?

> Who could get to know the heavens unless by heaven's gift, and discover the god unless he himself were part of the gods? Who could discern and encompass in his narrow breast the extensive mass of this vault which opens up without end, the dances of the stars and the blazing roof of heaven and the eternal war of the planets as they move against the constellations, [and the land and the sea beneath the sky, and what lies beneath each], unless nature had given sacred vision to our minds, turned to herself a kindred intelligence and communicated so great a science? Who, unless there came from heaven a power which calls us skywards to the sacred exchanges of things?

But this reassuring sense of cognitive solidarity between human being and divine creator is abruptly cut short by the lines that follow (2.136–44):

> haec ego divino cupiam cum ad sidera flatu
> ferre, nec in turba **nec turbae** carmina condam
> sed solus, vacuo veluti vectatus in orbe
> liber agam currus non occursantibus ullis
> nec per iter socios commune regentibus actus,
> sed caelo noscenda canam, mirantibus astris
> et gaudente sui mundo per carmina vatis,
> **vel quibus illa sacros non invidere meatus**
> **notitiamque sui, minima est quae turba per orbem.**

> This is the theme that I wish to carry to the stars with divinely inspired breath. I will not compose my song in the crowd **nor for the crowd**, but I will go alone as if, carried round the empty circuit, I were driving my chariot freely with no one obstructing me or steering a course beside me over common tracks. And I will sing a song to be recognised by heaven, as the stars marvel and the heaven rejoices in the song of its own bard, **or by those (mortals) whom the stars themselves do not begrudge knowledge of themselves and their sacred journeys: this is a very small crowd throughout the world.**

Line 137 relates to the exclusivity of both the poet and his target audience, to which 138–40 and 141–4 provide respective elaboration.

To concentrate specifically on the nature of the target audience, we appear at first glance to be faced with a contradiction, one of many in the poem: how can the heavenly elements both imbue mortals with and begrudge them knowledge of the stars? Why is it that the number of successful students of astrology is likely to be very small (2.144 *minima . . . turba*) when it has been previously implied that the god grants *all* humans the capacity to comprehend the stars (cf. 2.105–8)?

It might be attractive to think of this not as a contradiction but as a clarification of the relationship between god, human and intellect, and one in line with Stoic philosophy. One could argue that, although the universe does want human beings to understand its workings, and although human beings are endowed alike with a divinely given *capacity* for intellect and wisdom, it is up to the individual to *exercise* this capacity and apply reason in order to distinguish truths from falsehoods: this cerebral intensity thus explains the exclusivity of the successful group.[30] Such an interpretation would have the additional benefit of preventing the didactic project from imploding at this stage: a situation whereby individuals are required to work hard to apply reason at least retains a role for the teacher of astrology, who might otherwise have been rendered redundant if knowledge of the cosmos were viewed as an innate gift from a deity.

But even this conciliatory philosophical approach cannot withstand the sentiment of the lines that follow (2.145–9):

> illa frequens, quae divitias, quae diligit aurum,
> imperia et fasces mollemque per otia luxum
> et blandis diversa sonis dulcemque per aures
> affectum, ut modico noscenda ad fata labore.
> **hoc quoque fatorum est, legem perdiscere fati.**

> Vast is that crowd which esteems riches and gold, power and the rods of political office, and soft luxury amidst leisure, and the diversions of coaxing sounds and a sweet feeling through the ears—how modest the labour compared with an understanding of the Fates. **For this is also controlled by the Fates: to learn thoroughly the laws of Fate.**

As with 2.105–25, we are invited to accept a notion that human intellect and an ability to learn is preordained; but, unlike the positive noises of inclusivity in the previous passage, there is in these lines a

[30] See Mann (2011).

strong sense of absolute exclusivity: the Fates have already decided who will learn about their laws. From this point, Manilius starts his astrological instruction, on the different levels of diversity among the signs of the zodiac and the complex relational patterns between them (2.150ff.). But one is entitled to ask to what ends the teacher offers this instruction. If one tries to reconcile all that the poet has said in 2.105–49, one reaches the awkward conclusion that only a small group of individuals, those who make best use of their innate intellect, will come to understand the workings of the cosmos, and that the Fates have already decided who those individuals are. The role of the teacher now seems decidedly weak, as Manilius has set in motion one of the underlying tensions in the poem, between strong determinism and the ability of a teacher to influence students and, thus, become instrumental in changing the course of those future events.[31] The prospects for the student appear even more grim, as he has no way of knowing whether he is fated to be one of the chosen few, and must proceed in the blind faith that his mental effort is (fated to be) rewarded.[32]

Specific insight into Manilius' pedagogical strategies is offered towards the end of Book 2 (2.749–54):

> undique miscenda est ratio per quam omnia constant.
> verum haec posterius proprio cuncta ordine reddam;
> nunc satis est docuisse suos ignota per usus,
> ut, cum perceptis steterit fiducia membris,
> sic totum corpus facili ratione notetur
> et bene de summa veniat post singula carmen.

> From all quarters must be pieced together the rationale by which all things are ordered. But I will return to all these things later in their own proper order. For now, it is enough to have taught things unknown through their uses, so that, once faith has been established through an understanding of the parts, the entire body may thus be comprehended through simple reasoning/calculation, and my song may progress fittingly from discussion of individual details to the whole.

Manilius, then, professes a pedagogical creed that seeks to ensure that the student masters individual details before being introduced to the

[31] For useful discussions of the tension between teaching and determinism in the poem, see Neuburg (1993) 257–82, esp. 276–81; Volk (2002) 202–9.

[32] See further Neuburg (1993) 263–4.

ways in which these elements interact with each other to produce an understanding of the whole.[33] The teacher justifies this approach by two fine analogies: teaching children how to write (2.755–60) and building a city (2.772–83). He rounds off the section with the following statement of intent (2.784–7):

> sic mihi conanti tantae succedere moli
> materies primum rerum, **ratione remota**,
> tradenda est, ratio sit ne post irrita neve
> argumenta novis stupeant nascentia rebus.

> So, as I attempt to perform so great a task, I must first set out the material of things, **with the reasoning/calculation withheld**, so that hereafter the reasoning might not be given in vain, and that my arguments might not be struck numb at the outset by some new things.

At this stage of the instruction, then, the student has been made aware that *ratio*—a term which can encompass logic, reason and calculation—will be the key to success in understanding the intricacies of astrology: but at least some of that *ratio* is withheld until the teacher can be sure that the student has understood the basic elements. This is a fair enough position for a teacher to adopt, perhaps, but it is then justified to ask at what stage the student will receive this additional information to enable him to appreciate the whole. The answer is . . . never. As will become apparent on a second reading of the poem,[34] it is particularly ironic that the explanation of pedagogical strategy should follow straight on from an abortive section on the planets (2.738–48). The success of casting a horoscope rests on an ability to appreciate the complex interactions between the signs of the zodiac, their *paranatellonta* (the constellations that rise simultaneously with the zodiacal signs) and the planets. And yet it is specific information about the planets which is lacking in the poem: though their importance to astrology is regularly drawn to the student's attention, any

[33] This general creed is articulated more briefly elsewhere; cf. 2.761–71, 3.158–9, 3.586–9 (see also discussion of these passages later). A similar teaching strategy is advocated by Manilius' successor in astrological teaching, Firmicus Maternus (e.g. *Math.* 2 *praef.* 1, 2.14.1, 2.20.14, 3.1.19).

[34] At this point, it might be appropriate to refine slightly my position on the slippage between internal student (Reader) and external reader (reader). Whereas Reader and reader might be very closely connected on a first reading of the poem, a second reading would allow the reader to distance himself from the Reader: the reader would be able to smile at the credulous Reader taken in by Manilius' disingenuous didactic creed.

discussion of the planets is abortive and perpetually deferred to a later stage.[35] Many have noticed the glaring omission, and various explanations have been offered.[36] I am inclined to adopt a different approach on this matter, in that I suspect such 'forward glances' are part of a rhetorical strategy to keep the student engaged in the lesson to the very end before realising that he has been disappointed. Similarly deceptive strategies can be detected in the forward references in Ovid's *Fasti*, a work and author contemporaneous with Manilius.[37]

But it is not just specific elements of astrology, like the planets, which are ultimately missing from the lesson: again, as a second reading of the poem would make clear, Manilius in fact makes little attempt in the extant poem to relate each piece of information to the other to enable the student to appreciate the whole. As Volk (2009: 125) neatly puts it, 'what we are presented with in the *Astronomica* is thus an astrological alphabet but not a text, a storehouse of building materials but not a structure'.

By the end of Book 2, then, the student might be justifiably confused, if not despondent. He has no way of knowing whether he is fated to be able to learn the complex art of astrology. And even if he proceeds on the assumption that he is one of the chosen few, an expectation of integrated astrological knowledge is set up which will never be realised. The student will ultimately discover that he is to be perpetually denied the ability to cast his own horoscope and will be left to wonder about what might have been. This will not say much for the overall learning trajectory that Manilius envisages for his student: the student may be left to conclude that the teacher did not think the

[35] For the professed importance of the planets, cf. e.g. 1.15, 51–2, 670–1, 2.643–4, 3.96–155 *passim*. For specific deferral of the topic, cf. 2.961–7 esp. 965 *haec mihi sub certa stellarum parte canentur*, 3.581–9.

[36] Discussion of the planets may have been contained in a now lost section of the work, perhaps in the lacuna of Book 5 (on which see later discussion); perhaps it is a matter of generic affiliation and Stoic preference, since the father figure of astronomical poetry, Aratus, refuses to include discussion of the planets (Arat. 454–61), probably because they do not adhere to the determinability of his Stoic design for the universe (see e.g. Hunter (1995)). For detailed discussion of possible motivations behind the missing planets, and Manilius' avoidance of planetary influence in general, see Volk (2009) 48–57, 116–26.

[37] Though *Fasti* only consists of six books, treating the months of January to June, Ovid looks forward to handling topics and festivals in books on August (5.147–8) and December (3.55–8, 199–200). As I follow the scholarly consensus that the last six books of the poem were never written, such forward glances are disingenuous; for the deliberate intention to compose a six-book poem, see e.g. Feeney (1992), Newlands (1995) 124–45. See also Hutchinson (2008) 241.

student—*any* student—capable of going beyond the basic elements. But these are thoughts for the future, for at this stage the student carries on regardless . . . into the mathematical mire of Book 3.

CALCULATIONS AND COMPLICATIONS: BOOK 3

If Book 2 presented a greater challenge to the student than the first book, Book 3 increases the challenge further. The poet opens this book with due warning (3.31–42):

> at mihi per numeros ignotaque nomina rerum
> temporaque et varios casus momentaque mundi
> signorumque vices partesque in partibus ipsis
> luctandum est. quae nosse nimis, quid, dicere quantum est?
> carmine quid proprio? pedibus quid iungere certis?
> huc ades, o quicumque meis advertere coeptis
> aurem oculosque potes, veras et percipe voces.
> impendas animum; nec dulcia carmina quaeras:
> ornari res ipsa negat contenta doceri.
> et, siqua externa referentur nomina lingua,
> hoc operis, non vatis erit: non omnia flecti
> possunt, et propria melius sub voce notantur.

> But I must struggle through numbers and the unknown names of things, and the seasons, the shifting fortunes and movements of the universe, and the alternations of the constellations, and even the parts within these very parts. Indeed, how great a task is it to put into words what it is too much to know? Indeed, how great a task is it to speak it in fitting poetry and to yoke it to a regular metre? Come here, anyone who can turn both ears and eyes to my undertaking, and perceive words which are true. You must use up your mind. Nor must you look for a song that is sweet: the theme itself refuses to be decorated and is but content to be taught. Furthermore, if there are any names which are spoken in a foreign tongue, this is the responsibility of the theme rather than the poet. Not all names can be shaped (into the Latin language), and they are more appropriately expressed in their native tongue.

The general theme of Book 3, the measurement of time, will require extended mathematical calculations (31 *numeros*) and aspects of astrology for which a Latin translation is not possible (31, 41–2). This will be complicated, adds Manilius, by the poetic medium of the instruction, which will struggle to incorporate such prosaic subject matter and will leave little room for pleasant poetic embellishment (34–5, 38–9).[38] Such a momentous endeavour requires the student to apply all his senses, especially the mind (38 *impendas animum*), and the size and cerebral nature of the task is reiterated at regular intervals throughout the book.[39] In this way does the poet forewarn the student of the difficulties to come—difficulties which are potentially insurmountable (34 *quae nosse nimis*)?–as he quickly enters into the mathematical mire of what is a crucially important aspect of casting a horoscope, the ability to calculate the degree of the Ascendant (*horoskopos*) at the moment of birth (3.203–509).

But it is not just the mathematics which causes difficulties in Book 3, as the book itself plays host to several oddities which might only serve to confuse the student still further. The first topic to be handled is the circle of the twelve *athla* or 'lots' which govern different areas of a human's life (3.43–159). As scholars have duly noted, this appears to represent an (albeit similar but) alternative system of thought to the twelve temples (*dodecatropos*) just expounded at the end of Book 2 (856–967).[40] Indeed, the particular incompatibility within Manilius' poem of the discussion of the *athla* is demonstrated by some peculiar information (3.142–4, 149–55):

> non alia est sedes, **tempus**ve genusve **medendi** 142
> quae sibi deposcat vel **cuius tempore** praestet
> auxilium et vitae sucos miscere salubris. 144
>
> seu ferat officium nutus blanditus in omnis,
> aspera sive foro per litem iurgia temptet, 150
> fortunamve petat pelago ventisque sequatur,
> seu Cererem plena vincentem credita messe
> aut repetat Bacchum per pinguia musta fluentem,
> hac in parte **dies atque hac momenta dabuntur**,
> si bene convenient stellae per signa sequentes; 155
>
> No other seat is there (i.e. the eleventh lot) which claims for itself **the time for administering medicine** and the type of remedy to

[38] This stands in marked contrast to the famous 'sweet honey of the Muses' of Lucretius, where theme and medium are felt to complement one another (Lucr. 1.931–50, 4.1–25).

[39] Cf. 3.36–7, 43 *subtili . . . perspice cura*, 46 *vigilianti . . . sensu*, 186, 203 *agili . . . corde*, 211 *res est tam plena laboris*, 276 *animo . . . sagaci*, 295 *vigilanti . . . mente*, 618 *memori . . . mente*.

[40] For the unique complexity that Manilius gives to his astrological system here, see Volk (2009) 76–90.

administer, or **in whose time** help and the mixing of life-saving juices is at hand . . .

Whether one offers their service, showing flattery towards every nod, or tries their hand in the Forum at bitter disputes through litigation, whether one seeks their fortune on the sea and pursues it with the winds, or seeks a crop that outdoes the (seed) entrusted to it with a full harvest or seeks once again Bacchus flowing with rich must—in this part (i.e. the twelfth lot) **will the day be given and the very moment**, if the planets moving through the signs are favourably situated.

Elsewhere Manilius champions genethliological astrology and a belief in strong astral determinism at birth.[41] In these two passages, however, Manilius' advice moves into the realm of catarchic astrology, the branch of astrology which makes use of the position of the stars at a particular moment to judge whether or not it was favourable to conduct a particular activity.[42] The move is both sudden and brief, but enough potentially to destabilise the student.

Manilius, then, juxtaposes two different systems—the *dodecatropos* and the *athla*—without making any cross references between them. The student is left in a quandary: either both systems are valid, in which case astrology is (for the first time) revealed as a dynamic craft resistant to static rules;[43] or only one of the systems is 'correct', in which case the student is forced to choose in the full knowledge that he may be wrong.

From two apparently incompatible systems, we move to one rule (unhelpfully) masquerading as two. In discussing the methods for calculating the degree of the Ascendant (3.203–509), Manilius presents the student with two options: either assign a rising time of two hours to each sign of the zodiac (3.218–46) or assign fifteen degrees of the zodiac to each hour that passes (3.483–509). As scholars have rightly noted, these two methods represent two sides of the same coin: if fifteen degrees are assigned to every hour, then every sign of the zodiac, occupying thirty degrees, takes two hours—exactly what the first method had expounded.[44] Whether or not one regards this as a 'blatant self-contradiction' (Volk 2011b: 105), it is certainly an unhelpful

[41] See, most directly, 4.14–22, 98–101. [42] See Goold (1997) lxiv.

[43] See Barton (1994a) 92–4; Volk (2009) 81–2.

[44] For discussion of and general dismay towards this Manilian complication, see famously Housman (1903–30) 3.xxi; also Brind'Amour (1983); Volk (2009) 92–3; Volk (2011b) 105–6.

complication to a book which seems intent on creating difficulty for the student beyond the challenges of the subject matter itself.

One final sequence may serve as illustration of the potential frustration faced by the student in Book 3. This is the advice the poet offers on determining the length of a mortal's life (3.560–617). As will become clear in a later chapter, the ability to use the stars to predict the length of life—or, in other words, the time of death—was an issue both fascinating and highly contentious in contemporary Roman society.[45] And yet the poet pays no lip service to the politically contentious nature of the information he purports to be about to divulge. The poet's political complacency might be justified, however, when one considers what he actually does say/teach. The lesson starts promisingly enough, as the teacher sets out the number of years conferred on mortals by each sign (3.567–80). This is concise and, apart from some poetic periphrasis, easy enough to comprehend. Any sense of mental accomplishment on the student's part is, however, soon undermined by another piece of Manilian didactic retuning (3.581–5):

> nec satis est annos signorum noscere certos,
> ne lateat ratio finem quaerentibus aevi:
> templa quoque et partes caeli sua munera norunt
> et proprias tribuunt certo discrimine summas,
> cum bene constiterit stellarum conditus ordo.

> Nor is it enough to know the fixed number of years given by the signs, lest the reasoning lay hidden from those seeking to know the measure of a person's life: the temples and parts of the sky also know their own gifts and they impart their own amounts with precise distinction, when the established order of the planets stands by favourably.

First, the negatively phrased opening *nec satis* only serves to emphasise the insufficiency of the struggling student's knowledge so far.[46]

[45] The imperial sensitivity surrounding this issue resulted in the aged Emperor Augustus banning all astrological consultations of this type; see Chapter 5.

[46] There is in this regard a marked contrast between the styles of Manilius and Lucretius, both exponents of Roman philosophical didactic. Manilius is not opposed to opening new instruction with (the demoralising) *nec/non satis*—cf. also 2.270, 2.331, 2.643, 4.409—a phrase which Lucretius never uses. Lucretius prefers to build up the confidence of his student by stressing how forthcoming information draws directly on previous instruction, with regular phrases such as *(ut) (supra) docui* and *paulo ostendimus/diximus ante*; see Bailey (1947) 34, 160–1, 163–5; Wiltshire (1974); Clay (1983) 35–40, 181–5; Volk (2002) 75–7; Hutchinson (2008) 232–5. There are no phrases of this kind to comfort the Manilian student; see also Hutchinson (2008) 240–2, who notes the marked absence of motifs of structural cohesion in Manilius compared to his fellow Roman didactic poets.

The knowledge thus far acquired needs to be complemented by an understanding of other heavenly elements, the temples, which affect the number of years given to mankind in conjunction with the favourable disposition of the planets (3.583–5). Unfortunately, Manilius is once again only prepared to reveal some of the relevant information at this stage—the temples (3.590–617)—and once again it is the planetary influence which is deferred to a later stage (3.586–9):

> sed mihi templorum tantum nunc iura canentur;
> mox veniet mixtura suis cum viribus omnis,
> cum bene materies steterit percognita rerum
> non interpositis turbatarum undique membris.

> But for now I will only sing about the laws of the temples; in due course will come the whole mixture of elements with its full force, as soon as the constituent portions of things have been thoroughly learnt and stand firmly in mind, and are not thrown into confusion with parts scattered everywhere.

By now, it is a familiar tactic of Manilius' to follow emphasis on the importance of the planets with a decision to delay specific discussion. We saw it in Book 2, and it occurs earlier in this book during the discussion of the *athla*, where the need for favourable planetary disposition (3.155) is likewise met with delay on further elaboration (3.156–9):

> quarum ego posterius vires in utrumque valentis
> ordine sub certo reddam, cum pandere earum
> incipiam effectus. nunc, ne permixta legentem
> confundant, nudis satis est insistere membris.

> I will return later in due order to the power that these (planets) exert for good or ill, when I begin to expound their influence. For now, so that a mixture of things might not confuse the reader, it is enough to press on with the bare parts.

Ever thwarted is the student of astrology in Manilius, and Book 3 represents the lesson at its most intricate. The mathematical subject matter would have been challenging enough; but the student also has to contend with a teacher who presents incompatible systems, complicates rules and builds up the student's confidence with a promise of useful information only to be denied essential elements that render it (forever) incomplete. What the student is thinking at this moment we can only imagine, as he continues to remain silent. But the student's breaking point is close at hand.

THE STUDENT'S VOICE: DARKNESS AND DESPONDENCY IN BOOK 4

After the numerous complexities of Book 3, the student is desperately in need of some respite. And thankfully, following a forceful endorsement of determinism (4.1–118), this is precisely what the student appears to get during Manilius' discussion of the attributes, skills and associated occupations imparted by each of the signs of the zodiac to those born under them (4.122–291). In what follows, I maintain that Manilius' exposition here marks a striking departure from his didactic strategies earlier in the poem, in that he goes out of his way to present the student with something both extremely relevant and straightforward to comprehend.

First and foremost, Manilius forges very strong and simple correlations between the nature of the sign and the qualities/occupations assigned to its natives: the Ram gives birth to different kinds of wool workers (4.124–39); the Bull produces hardworking land-workers (4.140–51); the savage Lion and Scorpion produce warriors and those who delight in slaughter (4.176–88, 217–29); Erigone (Libra) gives birth to those who enjoy working with measures (4.189–202); the Centaur (Sagittarius) engenders keen horsemen and masters over beasts (4.230–42); and those born under either Aquarius or Pisces have affinity with water and its varied crafts (4.259–91). Even some of the less obvious characteristics attached to natives of a particular constellation are made to connect with either a discernible feature of the entity behind the sign or the time of year which it governs: so the oscillating financial fortunes of the Arian relate to the Ram's lot of growing and then losing (through shearing) a rich fleece (4.124–7); the sterility of those born under Erigone relates to her status as a virgin (4.202); the winter sign of Capricorn, which ushers in the new year, helps to explain its natives' penchant for warm clothing and changeable disposition (4.252–7); the overflowing urn of the Water Carrier means that its natives need not worry about accumulating wealth (4.272).

That these were some of the easiest concepts of astrology for the Roman layman to comprehend can be inferred from an interesting passage in Petronius' *Satyricon*, in which the pompous dinner host Trimalchio serves up an elaborate dish which, in the form of a rebus, alludes to the twelve signs of the zodiac (Petr. 35.1–5):[47]

> laudationem ferculum est insecutum plane non pro expectatione magnum, novitas tamen omnium convertit oculos. rotundum enim repositorium duodecim habebat signa in orbe disposita, super quae proprium convenientemque materiae structor imposuerat cibum: super arietem cicer arietinum, super taurum bubulae frustum, super geminos testiculos ac rienes, super cancrum coronam, super leonem ficum Africanam, super virginem steriliculam, super libram stateram in cuius altera parte scriblita erat, in altera placenta, super scorpionem . . . [pisciculum marinum], super sagittarium oclopetam, super capricornum locustam marinam, super aquarium anserem, super pisces duos mullos. in medio autem caespes cum herbis excisus favum sustinebat.

> After our applause, there followed a course which was evidently not as grand as we had expected; nevertheless its novelty turned the eyes of everyone. For placed on a round tray were the twelve signs (of the zodiac) arranged in a circle, above which our creator of matter had placed an appropriate food consistent with the relevant sign: above the Ram there was ram's head chick pea; above the Bull a slice of beef; above the Twins a pair of testicles and kidneys; above the Crab a garland; above the Lion an African fig; above the Virgin a barren sow's womb; above the Balance a pair of scales with a tart on one pan and a cake on the other; above the Scorpion [a little fish]; above the Archer a raven; above Capricorn a lobster; above the Water Carrier a goose; above the Fishes two mullets. And in the middle a piece of cut grassy turf was supporting a honeycomb.

The rebus works at a basic level, often by simple association. Only at times is slightly more thought required: so the Lion is connected to the fig through its African origin, the Archer and the raven are connected (most likely) by the sharpness of their extremity (arrow tip/beak), and the goose is connected to Aquarius through its love of water.[48] What

[47] The text of Petronius is taken from Mueller (1995). The text relating to the signs of Scorpio, Sagittarius and Capricorn in this passage is not unproblematic; see Rose and Sullivan (1968); Smith (1975) 75–6; Schmeling (2011) ad loc. For the most detailed study of Petronius' astrology here, see de Vreese (1927).

[48] It is only the pairing of the Crab with the garland which remains more elusive at this stage. It may be an astronomical allusion to the fact that, as constellations (Cancer and Corona), one of them is setting at the same time as the other is rising; see Schmeling (2011) 129. Then again, Trimalchio will later explain that he has made an exception of Cancer through fear of causing damage to his own natal sign (Petr. 39.8), a sentiment which directly ties this episode into Trimalchio's more general obsession with the Fates and death; see Grondona (1980) 9–75, esp. 17–25.

is important here is that the rebus appears to be perfectly comprehensible to the guests, even to the narrator Encolpius, who is elsewhere susceptible to confusion.[49] Not that this prevents Trimalchio from using the dish as an excuse for a pompous display of his astrological knowledge (Petr. 39.5–15):

> 'caelus hic, in quo duodecim dii habitant, in totidem se figuras convertit, et modo fit aries. itaque quisquis nascitur illo signo, multa pecora habet, multum lanae, caput praeterea durum, frontem expudoratam, cornum acutum. plurimi hoc signo scholastici nascuntur et arietilli.' laudamus urbanitatem mathematici; itaque adiecit: 'deinde totus caelus taurulus fit. itaque tunc calcitrosi nascuntur et bubulci et qui se ipsi pascunt. in geminis autem nascuntur bigae et boves et colei et qui utrosque parietes linunt. in cancro ego natus sum. ideo multis pedibus sto, et in mari et in terra multa possideo; nam cancer et hoc et illoc quadrat. et ideo iam dudum nihil supra illum posui, ne genesim meam premerem. in leone cataphagae nascuntur et imperiosi; in virgine mulierosi et fugitivi et compediti; in libra laniones et unguentarii et quicumque aliquid expendunt; in scorpione venenarii et percussores; in sagittario strabones, qui holera spectant, lardum tollunt; in capricorno aerumnosi, quibus prae mala sua cornua nascuntur; in aquario copones et cucurbitae; in piscibus obsonatores et rhetores. sic orbis vertitur tamquam mola, et semper aliquid mali facit, ut homines aut nascantur aut pereant. quod autem in medio caespitem videtis et super caespitem favum, nihil sine ratione facio. terra mater est in medio quasi ovum corrotundata, et omnia bona in se habet tamquam favus.'

> 'This heaven, in which the twelve gods reside, turns itself into just as many figures, and in this way becomes the Ram. And so anyone born under this sign owns many heads of cattle, much wool, a hard head besides, a brassy front and a sharp horn. Very many rhetoricians are born under this sign, as are ram-like individuals.' We praise the wit of our astrologer, and so he continued: 'Then the whole heaven becomes a little Bull. And so at that time are born stubborn men and ploughmen and those who can feed themselves from their own means. But under the Twins are born pairs-in-hand, oxen, lecherous men and those who befoul each orifice. Under the Crab was I born: and so I stand with many feet, and I possess many things from both land and sea, for the crab is at home in both environments. It is for this reason a little earlier that I put nothing above his sign, lest I should weigh down my own natal sign. Under the Lion are born gluttons and domineering individuals; under

[49] Cf. Petr. 40–1, where Encolpius needs to ask a guest about the significance of the boar that wears a freedman's cap.

> the Virgin, womanisers, fugitives and those destined for prison; under Libra, butchers, perfume-sellers and those who weigh up anything; under the Scorpion, poisoners and assassins; under the Archer, boss-eyed individuals who look at the vegetables but take the bacon; under Capricorn, people in trouble who sprout horns through their misfortunes; under the Water Carrier, innkeepers and pumpkin-heads; under the Fishes, fish-fryers and orators. In this way does the world turn like a millstone and is always up to some mischief such that men either come into this world or die. But as for what you see in the middle—the turf and the honeycomb above it—I do nothing without reason. It is mother Earth in the middle, rounded like an egg, and containing within itself all good things like a honeycomb.'

Both the order in which Trimalchio presents his material (starting with Aries and progressing in order through the signs) and the characteristics he assigns to those born under specific signs chime with existing astrological texts, including Manilius, albeit with some personal embellishment.[50] So authentic does he sound that Encolpius readily (if amusingly) hails him as *mathematicus* (Petr. 39.6).[51] The implication of these sections is clear: even Trimalchio, who regularly displays errors of learning, particularly with regard to mythology,[52] has been able to comprehend the basic astrological correlations between sign and native to impress his audience; nor are any of his audience confused by the dish or its explanation. Ultimately, in his own satirical style, Petronius bears witness to the very basic level of knowledge that is required of a Roman to master the particular aspect of astrology that Manilius is espousing in 4.122–291.

But it is not just the simplicity of the instruction in 4.122–291 which provides welcome relief for the student. Just as importantly, the selectivity of the material provided by Manilius creates the

[50] See in particular the connection between Aries and wool (Petr. 39.5; Man. 4.124–34); Taurus and hard workers (Petr. 39.6; Man. 4.140–7); Gemini and love-making (Petr. 39.7; Man. 4.157); Cancer and riches (Petr. 39.8; Man. 4.165–72); Virgo and human coercion (Petr. 39.10; Man. 4.190); Libra and weights and measures (Petr. 39.10; Man. 4.203–8); Scorpio and murderous criminality (Petr. 39.11; Man. 4.217–22). See further de Vreese (1927).

[51] Other more subtle features contribute to an authoritative air to Trimalchio in this section. He is conversant in the technical term *genesis* to refer to his natal star (Petr. 39.8). Moreover, Trimalchio appears to take on the lofty position of Stoic creator of the universe when he is referred to as 'creator of matter' (*materiae structor*, Petr. 35.2), whose actions, in true Stoic fashion, are governed by 'nothing without reason' (*nihil sine ratione*, Petr. 39.14). See further Connors (1998) 112–14.

[52] See e.g. Rimell (2007) 122–3.

impression that he is directing his information specifically towards a Roman lay audience. On the few occasions when the poet draws attention to a particularly famous native of a specific sign, it is always chosen from an accessible and anecdotal Roman context: so the Bull gave birth to Serranus, Dentatus and (by inference) Cincinnatus (4.148–9), and Servius was born under Libra (4.213–14).[53] The lay focus of the information is evidenced by the fact that Manilius avoids discussion of occupations related to imperial government—in its political, financial, military and religious capacities—in favour of trades such as wool-working, metal-working, farming, sailing, shipping and commerce.[54] Manilius is being deliberately selective here, if one is justified to compare his list with the much fuller range of occupations found in the later astrological work of Firmicus Maternus (*Mathesis* 3–4), where there is a strong focus on political, military and religious duties.[55] As a consequence of his non-elite focus, the procurement of finance becomes a consistent reference point in Manilius' section.[56]

I am not suggesting that the craftsmen and tradesmen whose occupations are listed in this section would have been an intended readership for Manilius' work: the literacy skills required to comprehend

[53] All these figures are anecdotal exemplary figures accessible to the reader; cf. Cic. *Sen.* 56; Verg. *A.* 6.843–4; V. Max. 4.4.4–7; Col. *praef.* 13–14. The use of exclusively Roman exempla here is striking in view of Manilius' general marginalisation of Roman predecessors; see especially the list of literary predecessors at 2.1–48 and 3.5–26, where only 3.23–6 alludes to a Roman author (most likely Ennius).

[54] A full list of occupations mentioned in Manilius is given by Goold (1997) 378–9 in his index, under 'native, types of'. There are some particularly pertinent examples of Manilius' non-elite focus. Erigone confers rhetoric and a mastery of words (4.194), and yet Manilius avoids the obvious career paths of the orator and politician to focus on the stenographer (4.197–9). In similar fashion, the Libran invested with a talent for legal minutiae becomes not a lawyer or advocate but a man who provides private legal advice to the general populace (4.209–12). It is not surprising, therefore, that Manilius should avoid imperial exempla in favour of popular characters from the nostalgic Roman past.

[55] For religious personnel, cf. e.g. *Math.* 3.2.18, 3.3.6, 3.5.27; for military personnel, cf. e.g. *Math.* 3.2.20, 3.3.9, 3.4.8, 3.4.26, 3.5.34; for political personnel, cf. e.g. *Math.* 3.3.10, 3.4.29, 3.5.2, 3.5.14. For useful overviews, see Thorndike (1913) esp. 421–9; MacMullen (1971); Bram (1975) 315–22.

[56] Cf. e.g. 4.126 (*censusque beatos*), 129 (*quaestus*), 132 (*in quaestum vendere*), 166 (*varios quaestus artemque lucrorum*), 171–2 (*novas . . . praedas . . . pretio subitos . . . census*), 186 (*lucrentur*); 247 (*opes*). For this reason it is registered with surprise when financial gain is not the primary motive; cf. 4.192 *nec tam compendia census*; 222 *nec praeda;* 272.

our author would, in all likelihood, have been beyond them.[57] What I am suggesting, however, is that Manilius' relaying of non-technical, commonsensical information, related to the diverse social context at Rome, helps again to establish a rhetoric of inclusion which draws in a Roman reader from outside the upper echelons of imperial Roman society.

But once again, any feeling of comfort on the student's part is short-lived, as the teacher applies a layer of difficulty onto this apparently straightforward information with his discussion of the decans, the units of ten degrees of which there are three in any sign and 36 in the entire zodiac (4.294–386). Talk of decans takes the student away from the familiarly Roman back to the obscurely Greek world of geometry (4.298 *quam partem Graiae dixere decanica gentes*), a world in which one must once again exercise sharpness of mind (4.308–9, 368). Indeed, the decans within a particular sign have a profound effect on the influence that a particular sign exerts, and Manilius provides a summary of the implications of this new information on the student's learning (4.380–6):

> nec tantum lanas Aries nec Taurus aratra
> nec Gemini Musas nec merces Cancer amabit,
> nec Leo venator veniet nec Virgo magistra,
> mensuris aut Libra potens aut Scorpios armis
> Centaurusque feris, igni Capricornus et undis
> ipse suis Iuvenis geminique per aequora Pisces;
> mixta sed in pluris sociantur sidera vires.

> Not only will Aries take pleasure in wool, nor Taurus in the plough, nor the Twins in the Muses, nor the Crab in trade, nor only will the Lion come forth as a huntsman, nor the Maiden as a teacher, nor will the Balance have power only over the measures, nor the Scorpion over arms and the Centaur over wild beasts, Capricorn over fire and the Youth himself over his own waters or the twin Fishes over the sea: but the stars are mixed up and become associated with more powers.

[57] Levels of literacy in the ancient world remain a contentious scholarly topic, and it would be inappropriate to project modern concepts of class and education back into the ancient world. But there remains a general impression that a complex hexameter poem would have required an advanced level of literacy reserved for the upper classes of society; see esp. Harris (1989) 175–284, esp. 222–9; Bowman (1991).

Despite the fact that he is asking the student to *supplement* information on the decans to the initial simplistic discussion of the influence of the signs, Manilius' phraseology here is noticeably harsh and off-putting for the struggling student: with its relentless repetition of negative connectives (*nec*) and its insistence on mentioning every sign of the zodiac, these lines read like a deliberate and systematic undoing of all the knowledge the student thought he had previously grasped in 4.122–291.

At this point, there is a crucial new development: the voice of the didactic student, hitherto silent, is directly evoked in a cry of despair (4.387–9):[58]

> 'multum' inquis 'tenuemque iubes me ferre laborem,
> **rursus et** in magna mergis caligine mentem,
> cernere cum facili lucem ratione viderer.'

> 'Extensive', you say 'and subtle is the toil you bid me undergo. **Once again**, indeed, you are plunging my mind into great darkness, just when I was appearing to see the light with some easy reasoning.'

Scholars have paid insufficient attention to the significance of this intervention. Didactic poets typically construct their students as tacit recipients of instruction. This is not to deny that, in some cases, the student's active presence is detected when the teacher appears to respond to (unquoted) questions posed by the student.[59] However, whether we are meant to interpret these student questions as having been posed in advance or during the course of the teacher's instruction, the important thing to note is that the student's actual voice remains undocumented in the text: the teacher is the master of ceremonies in

[58] The student's words have a distinctly Virgilian ring. Their structure recalls Aeneas' impassioned opening words to Dido before retelling his woes at Troy; cf. Verg. *A.* 2.3 *infandum, regina, iubes renovare dolorem.* Moreover, as noted by Feraboli, Flores and Scarcia (1996–2001: II.361), the student's acknowledgement that his toil is both extensive (*multum*) and subtle (*tenuem*) recalls Virgil's *Georgics*, whose teacher had described his agricultural instruction in similar fashion; cf. Verg. *G.* 1.176–7 *possum multa tibi veterum praecepta referre,/ni refugis tenuisque piget cognoscere curas.*

[59] Cf. e.g. Man. 2.433 *his animadversis rebus, quae proxima cura?* ('once these things have been noticed, what is the next concern?'), where the question may be imagined as coming from the student; see Neuburg (1993) 271. For the frequent pose of the didactic teacher responding to student questions, typically introduced by an appropriate form of *quaero*, cf. Man. 3.203–4 *forsitan et quaeras, agili rem corde notandam,/qua ratione queas*; also e.g. Lucr. 5.1091; Verg. *G.* 2.288; Ov. *Med.* 79; *Ars* 1.49; 1.375, 3.750; *Rem.* 487, 803.

his own didactic text, and he is not interrupted once he has started to speak.[60] Only on two occasions prior to Manilius is this didactic protocol compromised by having the words of the questioning student directly quoted. Both examples occur in the first book of Lucretius' *De Rerum Natura* (1.803–11, 897–903):[61]

'at manifesta palam res indicat' inquis 'in auras
aëris e terra res omnis crescere alique;
et nisi tempestas indulget tempore fausto
imbribus, ut tabe nimborum arbusta vacillent,
solque sua pro parte fovet tribuitque calorem,
crescere non possint fruges arbusta animantis.'
scilicet et nisi nos cibus aridus et tener umor
adiuvet, amisso iam corpore vita quoque omnis
omnibus e nervis atque ossibus exsoluatur;

'But plain fact shows openly', you say, 'that into the winds of the air out of the earth all things grow and are nourished; and unless the season gives way to the rains at a favourable time, so that the strawberry-trees sway under the melting of the clouds, and unless the sun nurtures them and offers its warmth of its own accord, nothing can grow—fruit, trees or animals.' True, and unless dry food and soft water come to our aid, we would at once lose our body, and all life would also be released from all our sinews and bones.

'at saepe in magnis fit montibus' inquis 'ut altis
arboribus vicina cacumina summa terantur
inter se validis facere id cogentibus austris,
donec flammai fulserunt flore coorto.'
scilicet et non est lignis tamen insitus ignis,
verum semina sunt ardoris multa, terendo
quae cum confluxere, creant incendia silvis.

'But often it happens on big mountains', you say, 'that the top branches of the high trees, being close together, are rubbed one against another when the strong South wind compels them to do so, until they burn bright when the flower of flame breaks out.'

[60] This is a defining feature of the didactic genre according to Diomedes, as it is poetry 'in which the poet himself speaks without interruption from any other person' (*in quo poeta ipse loquitur sine ullius personae interlocutione*, Diom. *Gramm. Lat.* 1.482.21 Keil).

[61] The text of Lucretius is taken from Martin (1953). I leave out of the discussion the intervention by female students at Ov. *Ars* 3.7–8 on the grounds that this occurs at an interval between instructional sessions: it is a rallying call for a new type of instruction at a point when the poet has just finished his (first) instruction.

> True, and yet fire is not implanted in the wood, but there are many seeds of heat which, when they stream together by rubbing, create fires in the woods.

A comparison between the Lucretian and Manilian examples of student intervention proves very instructive in highlighting the difference in student–teacher relationship in and didactic trajectory of each work. In both the Lucretian examples, the exchange takes place in a spirit of intellectual curiosity and debate. First, the student makes a valid point based on autopsy—an observation about plants (1.803–8) and trees (1.897–900)—which is used to draw an inference: respectively, that plant life contains the four elements and that wood contains flame. In both cases, the teacher immediately responds cordially by respecting the truth of the observation—with the encouraging *scilicet* (1.809, 901)—before informing the student that the inference drawn is incorrect (1.809–13, 901–3). The Lucretian teacher's relationship with his student, at least as far as these examples are concerned, suggests warmth, mutual intellectual respect and a desire for progressing the student's learning.[62]

The Manilian exchange, I would argue, strikes a very different tone. In the light of our discussion of the previous books, it would be quite wrong to interpret the student's words at 4.387–9 as a momentary loss of clarity. On the contrary, they reveal first-hand the student's own reflection on the entire learning experience since the beginning of the poem. This is the first time that the student has been able to grasp something truly astrological in the work and, especially as it comes after the intricacies of Book 3, it is all the more distressing when that long-awaited sense of surety is suddenly taken away: the student could see for but a brief time before he was plunged, once again (*rursus*, i.e. as in Books 2–3) into darkness. The student's words make a particularly telling contrast with the teacher's advice at 4.308–9:

> quae tibi non oculis, alta sed mente fuganda est
> caligo, penitusque deus, non fronte, notandus.

> This darkness must be put to flight not by your eyes but by deep contemplation, and it is in its deep interior, and not in its outward appearance, that the divine must be appreciated.

[62] For a similarly positive assessment of the teacher's relationship with his student in Lucretius, see Classen (1968); Wiltshire (1974); for a different view, however, see Mitsis (1993).

In much the same vein as Lucretius, Manilius sees his role as dispeller of darkness with the 'light' of mental clarity.[63] Unlike in Lucretius, however, the Manilian student confirms in 4.387–9, even towards the end of the poem/lesson, that he is still in the very *caligo* from which the teacher was hoping to lead him in 4.308–9, and the student's use of the language of sight (4.389 *cernere . . . lucem . . . viderer*) suggests that he still has not progressed from the eyes to the mind. Put another way, the student has not progressed beyond the visual, Aratean frame of reference of Book 1: only when he has a visual referent—such as a Ram and a Lion—can he make sense of the instruction.

Manilius' response to this student outburst (4.390–407) lacks the more personalised encouragement of Lucretius and amounts to little more than a cold rehash of Stoic doctrine espoused elsewhere in the work:[64] you are on a very difficult quest for god himself (4.390–4); only the opportunity to enquire, rather than success, is guaranteed (4.395), as man has to expend his very self in pursuit of astrological knowledge (4.406–7). Manilius follows this by applying yet more layers of difficulty to the instruction: taking on board all this information is *still* not enough (*nec satis est*, 4.409). In the course of Book 4, Manilius has moved from discussion of the signs (30°, 4.122–293), to the decans (10°, 4.294–386), to individual degrees (4.408–501), to changes within a degree (4.502–84). At each stage, the student does not see the complication coming, but is rather fooled into thinking that he knows the full picture, before successive layers of difficulty are applied. This will lead to a further outburst from the student towards the end of Book 4 (4.866–72):

> sed quid tam tenui prodest ratione nitentem
> scrutari mundum, si mens sua cuique repugnat
> spemque timor tollit prohibetque a limine caeli?
> 'conditur en' inquit 'vasto natura recessu
> mortalisque fugit visus et pectora nostra,
> nec prodesse potest quod fatis cuncta reguntur,
> cum fatum nulla possit ratione videri.'

[63] For Lucretius as bringer of light in the darkness, following on from the practice of Epicurus, cf. e.g. Lucr. 1.143–4 *dictis quibus et quo carmine demum/clara tuae possim praepandere lumina menti*, 1.146 *hunc igitur terrorem animi tenebrasque necessest* (= 2.59, 3.91, 6.39), 1.1114–17; for Epicurus, cf. 3.1–3, 5.10–12.

[64] I cannot agree with Neuburg (1993: 273–5) that Manilius' Stoic encouragement is a successful response to the flagging student.

> But what is the use in probing into the shining firmament with such fine reasoning, if anyone's own mind fights against it and if fear banishes hope and holds him back from the threshold of heaven? 'Look!', he says, 'nature is buried in a vast cavern and flees our mortal gaze and heart; nor can it be of use that all things are governed by the fates, since fate cannot be seen by any reasoning.'

This second outburst from the student has more serious implications than the first. For one thing, it is longer, as the student feels it necessary to cause greater interruption to the teacher to express his frustration: and note here that Manilius may have constructed this as a 'real' interruption, in that the words are uttered by the student (*inquit*, 4.869) rather than being mouthed by the teacher (*inquis*, 4.387).[65] The sentiment is also much more serious, as the student himself raises fundamental questions about the capacity of any mortal to comprehend the essence of an art which lies beyond his heart and mind—note the all-encompassing *cuique* (4.867) and *mortalisque . . . nostra* (4.870). Moreover, the language of sight reappears—*visus* (870), *videri* (872)—to reinforce the impression that the student has not progressed beyond the visual, Aratean mode of stellar enquiry from Book 1 and to suggest that, for the student, seeing is understanding. And reference to a student who is both without light and fearful (*timor*, 4.868) only continues Manilius' critical intertextual engagement with Lucretius.[66]

In effect, the student's learning experience so far has caused him to challenge, and implicitly reject, the Manilian notion that humans have a capacity for astrological knowledge. His words make a particularly pointed contrast with 2.105–8 (quoted earlier). Who can doubt a willing cognitive connection between heaven and man, asked Manilius? Well, the student can, and does so directly just two books later. It goes without saying that the student will not share Manilius' final assessment in the book: that reason triumphs over everything (4.932 *ratio omnia vincit*).

We are coming towards the end of Manilius' astrological lesson—four out of five books have been read—and yet there is little evidence of any progression in the student's competence in astrology. On the

[65] The text of 4.869 is problematic: *inquit* is Richard Bentley's conjecture, following on from *en*, the conjecture of Friedrich Jacob; both critics are trying to make better sense of the MSS *enim quid* or *enim quod*.

[66] For Lucretius, fear is the very thing that the philosopher teacher is charged with dispelling for the student; cf. esp. Lucr. 2.55–61 (= 3.87–93, 6.35–41).

contrary, the student's rejection of the basic rationale for learning, a rejection which the student nowhere takes back, marks a specific note of regression. This stands in contrast to the progression of the learner detectable in earlier didactic poems.[67]

DIGRESSION AND MISDIRECTION: THE FINAL PART OF THE LESSON IN BOOK 5

'Here would another poet have finished his journey' (5.1 *hic alius finisset iter*), says Manilius at the opening to Book 5. Any study of the didactic trajectory in this final book is complicated by the lacuna evident after line 709, a gap which can be filled with various possibilities (see later), and by the issue of whether or not this was indeed the last book of the poem. But by continuing to take a linear approach to the text, we may detect, as with the previous books, a similar sense of student frustration built into Manilius' didactic project.

A continuation of the opening lines of Book 5 hones right in on a central source of frustration for the student (5.1–7):

> hic alius finisset iter signisque relatis
> quis adversa meant stellarum numina quinque
> quadriiugis et Phoebus equis et Delia bigis
> non ultra struxisset opus, caeloque rediret
> ac per descensum medios percurreret ignes
> Saturni, Iovis et Martis Solisque, sub illis
> post Venerem et Maia natum te, Luna, vagantem.

> Here would another poet have finished his journey and, having dealt with the constellations against which move the divinities of the five planets and Phoebus with his team of four horses and Delia with her team of two, no further would he have crafted his work; rather he would return from heaven and in his descent would drive his chariot through the intervening fires of Saturn, Jupiter, Mars

[67] For the implied development of Perses from a lifestyle of injustice to justice in Hesiod's *Works and Days*, see Strauss Clay (1993). For the development of the student in Lucretius, see Clay (1983) 216–66. Even in Ovid's *Fasti*, a poem not strictly didactic but one that adopts regular didactic poses, progression can be detected in the figure of Ovid as interviewer, a figure who passes from naïve and tactless speaker to respectful and thoughtful aetiological enquirer; see Green (2001).

> and the Sun and beneath them you, o Moon, who wander after
> Venus and the son of Maia.

If the first line hints that this will be the final book, Manilius is fully aware of what is yet to be discussed—the planets—but successive sentiments continue to tease the reader in this regard. Lines 1–4 may give the impression that discussion of the planets is now at hand: any suggestion that *another* might conclude before discussing the *stellarum numina quinque* raises the expectation that Manilius will be different and will not avoid the planets any longer. But lines 4–7 clarify the author's position: the *alius* of line 1 is not someone who might omit discussion of the planets, but rather someone who might rush into a discussion of them prematurely. Manilius implies that he wishes to stick around in heaven before 'descending' into a discussion of the planets, and his intended subject for this book is revealed in the lines that follow (5.8–11):

> me properare etiam mundus iubet omnia circum
> sidera vectatum toto decurrere caelo,
> cum semel aetherios ausus conscendere currus
> summum contigerim sua per fastigia culmen.

> The universe also bids me to hasten to ride and journey across all
> the stars in the whole of heaven, now that I have once dared to
> mount my ethereal chariot and have touched the highest peak of
> the dome along its upward slopes.

For Book 5 then, it is a case of more stars, rather than planets. The stars in question turn out to be the *paranatellonta*, the extra-zodiacal constellations, as Manilius embarks on a lengthy description of the various skills and occupations imparted to humans by these constellations (5.32–709). In many respects this section works along the same lines as the discussion of zodiacal constellations at 4.122–291: colourful, wide-ranging, peppered with Roman anecdotal exempla and, most importantly, accessible in its underlying logic. So, for example, the Argo gives rise to captains and navigators (5.32–56); the Charioteer to horseman and chariot-riders (5.67–101); the Hare to quick-minded pursuits (5.157–73); the Arrow to javelin-throwers and archers (5.293–310); and the Lyre to musicians (5.324–38). Some signs require a little more thought from the student, but nothing too strenuous: for example, the scorching and canine qualities of the Little Dog Star (Canicula) give rise to unbridled spirits prone to anger (5.206–33); the Crown (Corona) gives rise to those who might enjoy arranging crowns/wreaths in a floral/garden context (5.251–69).[68]

However, there is one among these constellations, Andromeda, who is afforded special significance, as Manilius expands at some length on the mythological background to her catasterism (5.538–618), in what emerges as by far the longest 'digression' in the poem.[69] Manilius duly picks up the story at the point where Andromeda is being led to the sea shore to be sacrificed to the Sea Monster; Perseus chances upon the maiden, falls in love with her instantly and, after a lengthy battle with the Sea Monster, rescues Andromeda and takes her hand in marriage. While we should not underestimate the story's poetic and aesthetic charm, particularly in a work which can be dense in technical detail, extended (mythological) excursus in didactic poetry are generally felt to perform a practical function of aiding the instruction in some way.[70]

Does the Andromeda myth, then, provide any similar service for the student? In general terms, the situation looks promising. First, the digression is very well chosen, in as much as no other mythological episode involves so many future constellations interacting at the same time: Andromeda (e.g. 5.544), Perseus (e.g. 5.567), the Sea Monster—strictly Cetus (cf. 5.600), but often referred to in more generic terms during this story as *belua* (5.544, 608) and *monstrum* (5.581)[71]—Medusa's head (e.g. 5.567), and Andromeda's parents, Cepheus and Cassiopeia, knowledge of whose former impious transgressions is expected of the reader (5.540 *dirorum culpa parentum*). In effect, this myth is designed to explain the close proximity and eternalised poses of these very constellations in the night sky, as Manilius had outlined earlier in Book 1 (354–60):[72]

> . . . Cepheusque et Cassiepia
> in poenas resupina suas iuxtaque relictam
> Andromedan, vastos metuentem Pristis hiatus,

[68] See further Calcante (2002) 113–29.

[69] For recent studies of this episode, see Voss (1972); Murgatroyd (1994); DeNardis (2003) 104–5, 125–7; Uden (2011).

[70] For a summary of the complex didactic messages in the digressions of Ovid's *Ars Amatoria*, for example, see Green (2006) 16–18.

[71] For the inspiration behind Manilius' depiction of the Sea Monster, see Coleman (1983) and Lowe (2004).

[72] See also Goold (1959), who maintains that the myth may have originally arisen directly from the positioning of the stars.

[expositam ponto deflet scopulisque revinctam]
ni veterem Perseus caelo quoque servet amorem
auxilioque iuvet fugiendaque Gorgonis ora
sustineat spoliumque sibi pestemque videnti.

. . . and Cepheus and Cassiepia, bent backwards towards the punishment she caused, and beside her abandoned Andromeda, fearing the gaping jaws of the Sea Monster, [and she (Cassiepia) weeps for her, exposed to the sea and chained to the rocks], if Perseus were not maintaining even in heaven his former love and coming with aid and raising up the face of the Gorgon from which all should flee—a trophy for himself but a source of death for the beholder.

A strong sense of continuity is forged in Book 1 between the actions of these individuals both on earth and in the sky. The story in Book 5 offers a charming expansion of these endeavours on earth,[73] and this may cheer the student with the prospect of further continuity between earthly pursuits and heavenly qualities which, we may imply from his outburst in Book 4, was precisely the part of the Manilian instruction that he had found most accessible and easy to comprehend.

In fact, Manilius' intention to forge links between earthly and stellar activity may be a more embedded strategy within the story in Book 5 if one considers, literally, the trajectory of the action. When the battle between Perseus and the Sea Monster gets properly underway, it is striking to note how much emphasis Manilius places on the vertical movement of the two protagonists up and down, into and out of the water. The Sea Monster emerges headfirst from the water,[74] in response to which Perseus flies up to meet it.[75] The beast subsequently rises up to meet him, more fully out of the water, as the fight occurs on high,[76] and as often as it launches itself from the depths into the air, Perseus counteracts by rising higher.[77] Riddled with stab wounds, the

[73] Some variation, however, can be detected between the two versions. Most obviously, it is suggested in 1.358–60 that Perseus uses the Gorgon's head to defeat the Sea Monster, whereas in 5.538–611 Manilius concurs with Ovid (*Met.* 4.720, 727, 734) in having Perseus attack and defeat the Sea Monster with a sword; cf. Man. 5.594, 608; for these traditional variants to the myth, see Ogden (2008) 91–3. Furthermore, Uden (2011: 237–42) argues that the theme of love, specifically cited in 1.357 as the principal motivation for Perseus' action, is consciously downplayed in the story in Book 5.

[74] 5.581–6, esp. 581 *caput eminet undas.*

[75] 5.592–4, esp. 592 *subvolat.*

[76] 5.595–7, esp. 595 *subit*, 596 *alte*, 597 *sublimis.*

[77] 5.598–600, esp. 598 *subit, iaculata profundo*, 600 *subeuntis.*

Sea Monster finally sinks into the sea before rising dead to the surface,[78] after which Perseus bathes in the water before flying upwards to rescue Andromeda from the lofty rocks.[79] This sense of vertical movement is more pronounced than in the Ovidian version of the myth, where the Sea Monster's movement in and out of the water is mentioned only once (*Met.* 4.721–2), and its (more natural) horizontal trajectory is highlighted as it comes towards the coast like a ship (*Met.* 4.706–8).

Is this particular focus on vertical movement designed to look forward, in crude non-technical terms, to the rising and setting of their future stellar forms? The emphasis on movement in and out of the sea might be particularly interesting in this regard when one recalls the very popular and well-established poetic motif of expressing the rising and setting of constellations in terms of the animals/entities they represent plunging into and emerging from the sea. The motif goes back to Homer, who attests the notion that the stars set by bathing in Oceanus—the enormous river that encircles a flat-disced earth (*Il.* 18.607–8)—only to emerge again/rise more brightly.[80] Though Hesiod follows this Homeric lead on one occasion (*Op.* 564–7), he elsewhere uses πόντος (*Op.* 619–20), a generic term which invites slippage between Oceanus and any vast expanse of water. Subsequent authors, especially the Romans, are free to slip between the specific and the generic in this regard.[81]

Manilius is fully aware of this metaphor and alludes to the Homeric background (*Il.* 5.5–6) in his description of the Dog Star's brilliance (1.410–11):

> cetera vincuntur specie, nec clarius astrum
> tinguitur oceano caelumque revisit ab undis.

> The other stars are surpassed in beauty, and no brighter star is
> bathed in the ocean or revisits the sky from the waters.

[78] 5.608–11, esp. 608 *subsedit*, 609 *remeavit ad undas*.

[79] 5.612–15, esp. 613 *ex undis ad cautes pervolat altas*.

[80] Cf. *Il.* 5.5–6, 18.487–9 (= *Od.* 5.273–5, with Heubeck et al. (1988) ad loc.).

[81] For specific reference to Oceanus, cf. e.g. Arat. 593; Cat. 66.68; Verg. *G.* 2.481; *A.* 1.745; Prop. 4.4.64; Germ. *Arat.* 287, 396, 522, 569, 581, 590, 624. For a variety of generic expressions, see especially Ovid's *Fasti*; cf. 1.314 (*occiduas . . . aquas*), 1.457–8 (*aequora . . . patriis . . . vadis*), 2.73 (*Hesperias . . . undas*), 2.78, 4.164 (*viridis . . . aquas*), 4.388; see also Cic. *Arat.* fr. 33.381–2 (Soubiran); fr. 33.416 (Soubiran); Germ. *Arat.* 595, 605, 608, 613; Ov. *Tr.* 4.3.4.

His use of the metaphor is, however, nowhere near as widespread as one might expect, and in fact he reserves it (fittingly) for stellarised sea creatures, such as the Dolphin (1.346–7, 5.416–17):[82]

> tum quoque de ponto surgit Delphinus ad astra,
> oceani caelique decus, per utrumque sacratus.

> Then also does the Dolphin rise from the sea to the stars, the ornament of both the ocean and the sky, in each held sacred.

> caeruleus ponto cum se Delphinus in astra
> erigit et squamam stellis imitantibus exit . . .

> When the sea-blue Dolphin raises itself from the sea into the stars and emerges with stars that imitate scales . . .

In the light of Manilius' predilection for reserving the metaphor for sea creatures, as well as his apparent ease in slipping between the specific and the generic (there is no reference to Oceanus in the latter example), is the student being coaxed into interpreting the action of the Andromeda story in metaphorical terms? This comforting sense of slippage between the stellar and earthly worlds might also be felt in the depiction of Andromeda as an image of serenity and white radiance—a star before her time?—an image which stands in contrast to the popular tradition of the princess as dark-skinned.[83] Moreover, the delightful phrase *pontumque exstillat in astra*—5.604, 'and (the Sea Monster) sends in spray the ocean up to the stars'—captures vividly the merging of earthly and stellar worlds.[84] Finally, there is a subtle but noteworthy difference between the Manilian and Ovidian versions of the aftermath of the battle. Ovid's hero washes his hands with water drawn from the sea: *Met.* 4.740 *ipse manus hausta victrices abluit unda.* Manilius is more specific in having his hero bathe his whole body in the sea itself (5.612–13):

[82] This tactic is one that Manilius shares with Cicero, whose *Aratea* likewise reserves the motif for stellarised marine creatures; cf. Cic. *Arat.* fr. 33.381–2 (Soubiran) *mergitur unda/Delphinus*; fr. 33.416 (Soubiran) *illa [sc. Pistix] usque ad spinam mergens se caerula condit.* One may account for this narrowness of scope for the motif in terms of both authors' leanings towards a Stoic world of *sympatheia* between heaven and earth.

[83] The tradition of Andromeda's dark skin is alive in Manilius' own time: it is commented on in the *Epistula Sapphonis* (36 *Andromede patriae fusca colore suae*) and is implied in Ovid's comment at *Ars* 2.643 (*nec suus Andromedae color est obiectus*); see further Ogden (2008) 82–7; Lowe (2004) 147–8. The heroine's darkness for a quite separate reason is also attested in Accius' tragedy (*Andr.* frr. 70–1 Warmington), where Andromeda is seen in filthy disarray outside the Sea Monster's lair.

[84] The unique transitive use of *exstillo* in this passage should not be grounds for toning down the force of the phrase; for which see Jones (1984).

perfundit liquido Perseus in marmore corpus,
maior et ex undis ad cautas pervolat altas.

Perseus drenches his body in the liquid marble surface of the sea and, looking more magnificent, flies up from the waves to the high rocks.

This is no mundane detail. In emerging fully into the sea before resurfacing and, indeed, looking all the more magnificent for it (*maior*), Manilius' Perseus rehearses the very activity that his catasterised form will (metaphorically) perform each time it sets and rises.

To summarise to this point, the (poetically informed) reader/student will see in the Andromeda digression a pleasing sense of continuity between earthly and stellar activity. This, in turn, raises in the reader/student the expectation that the stellar gifts given to natives by the constellation of Andromeda will be consistent with the morals of the earthly story, as has been the case broadly speaking with the other *paranatellonta* discussed in Book 5. But Manilius' extended focus on Andromeda only makes more pronounced the surprise that follows: the constellation of Andromeda operates in a way very different from her character in the story (5.619–30). Far from inheriting any sympathy for imprisoned individuals, Andromedans will choose a career in the administration of severe justice in the form of the prison warden or executioner: 5.620–1 *immitis veniet poenaeque minister/carceris et duri custos*, 625 *carnificisque venit . . . imago*. Indeed, Manilius presses home the unfeeling nature of the Andromedan in this regard, an individual who is not only harsh (5.620 *immitis*, 621 *duri*) but also arrogant in his power and ready to profit financially from the misery of others (5.621 *superbe*, 625 *carnificisque . . . mortem vendentis*). And if we missed the central contrast between the Andromedan and the character of his mythical parent, Manilius confronts the contradiction directly (5.627–8):

. . . qui denique posset
pendentem e scopulis ipsam spectare puellam.

(the Andromedan is a man) who, in short, could have looked upon the girl herself (Andromeda) as she hangs from the rocks.

In other words, the Andromedan could take the position of Perseus in the myth (5.569–72), but could gaze upon the spectacle unmoved to action.[85] The fact that the Andromedan can *occasionally* act with

[85] The change in verb is telling: whereas Perseus merely sees Andromeda before resorting to action (5.569 *vidit*), Andromedans watch and take pleasure in the scene as if it were a spectacle for their amusement (*spectare*).

compassion to save a criminal (5.629–30, esp. 630 *interdum*) does little to mitigate the central contradiction in the native's character.

There is undoubtedly a tension here between the characteristics suggested by the myth of Andromeda and the lifestyles of the mortals born under her sign, a tension all the more striking and deliberate if Manilius had accessible to him an alternative tradition that the constellation of Andromeda gave rise (more naturally) to prisoners rather than prison guards.[86] Attempting to ease the tension, Hübner, a scholar expert in the intricacies of ancient astrology, points to the fact that the constellation of Andromeda appears in Pisces, a zodiacal sign which exhibits internal contradictions by virtue of the fact that its two Fishes are pointing in different directions.[87] Uden adopts a more holistic approach by taking on board four of the constellations involved in the story (Cepheus, Cassiopeia, Andromeda and Cetus) to make links between the mythical epyllion and what Manilius tells us about the lifestyles of those born under its varied protagonists.[88] For example, Cassiopeia's penchant for jewellery (5.504–8, 516–19) tellingly includes 'golden chains' (519 *aurea vincla*), a motif which re-emerges unexpectedly in the Andromeda epyllion in the form of the chains that bind the heroine to the rocks (5.551): the worlds of personal adornment and imprisonment by stones have become curiously merged. Here and elsewhere, Uden detects thematic and verbal resonances between the myth and the surrounding discussion of natal traits which draw attention to 'the predictive logic of astrology: oblique, riddling, enigmatic'.[89]

Interesting though these scholarly approaches are, they have the effect of downplaying Manilius' flirtation with opposites when he draws our attention to the central inconsistency in 5.627–8 (quoted earlier): Andromedans, says Manilius, will simply not relate sympathetically to Andromeda. I take Manilius' lead in maintaining a strong sense of contradiction here, and as such concur with Calcante, who draws particular attention to the pitiable emphasis on punishment and lamentation in the epyllion, the very things which the Andromedan native can look upon with indifference and indeed exacerbate.[90] Modern

[86] See Teucros in Boll (1903) 41–2 (lines 20–5).

[87] See Hübner (1984) 193–201; (2011) 149–52.

[88] See Uden (2011) 247–52.

[89] See Uden (2011) 247.

[90] See Calcante (2002) 129–33. But Calcante interprets the Andromeda epyllion in largely literary terms (parallelism through opposition) rather than as part of a didactic strategy of misdirection.

scholars are right to see the Andromeda myth as emblematic of the poem as a whole, in terms of its thematic tensions, generic affiliations and Stoic undercurrent.[91] But, as I have argued, the myth is also emblematic in terms of its role in facilitating complexity for the student and increasing his frustration.[92]

Manilius' discussion of the *paranatellonta* is interrupted at 709 by an apparent lacuna; the text resumes with the latter part of a discussion on the relative luminosity of the stars in the sky which develops, in the final lines of the poem, into a comparison with the Republic (5.710–45). The missing portion of text might have been quite substantial, potentially over 200 verses.[93] So what might it have included, beyond the end of the section on the extra-zodiacal constellations and the beginning of the section on stellar luminosity?[94] Goold maintains that the missing lines must have contained the currently elusive and much-promised discussion of the planets.[95] Internal evidence, however, suggests otherwise. As Volk astutely notes, the final excursus on the luminosity of the fixed stars strongly suggests that Manilius has not yet turned to the topic of the wandering planets, and I agree with her that a discussion of the *settings* of the *paranatellonta*, promised at 5.27–9, provides a more logical candidate for the missing section.[96] I would add, however, that it might only have taken a few (missing) lines for the student to have voiced further frustration about the complexity of the lesson. Just as in Book 4, where an outburst was triggered by the sudden complication of a seemingly easy concept, so in Book 5 the surprising complication caused by the Andromeda epyllion might have elicited a similar if not stronger reaction against the teacher and his didactic project.

[91] For thematic tensions, see Lowe (2004); for generic affiliation with elegy and epyllion, see Voss (1972) and Uden (2011), respectively; for Stoic undercurrent, see e.g. Uden (2011).

[92] It is at this point that Hübner's arguments (see n. 87) can be used to bolster the case: it may be the case that Manilius' 'logic' makes sense to the reader expert in astrology, but this is precisely *not* the target audience for the instruction.

[93] See Goold (1983) 66–7, who calculates that either 176 or 220 lines of text may be missing.

[94] For the missing part of the section on the *paranatellonta*, Goold's suggested reconstruction from Firmicus Maternus is probably the best that can be managed; see Goold (1997) xcvi–xcvii.

[95] See Goold (1954) 369–72; Goold (1997) xcvii–xcviii and esp. Goold (1983), where it is suggested that the section on the planets may have been deliberately removed by an eager astrologer: such a suggestion buys into an image of Manilius as an author of sincerity and clarity of instruction that I can support.

[96] See further Volk (2009) 119.

At any rate, the secrets of the planets, and hence a proper means of creating a horoscope, remain undisclosed in the extant text, a point towards which our author may be gesturing in the closing lines (5.740–5):

> sunt stellae procerum similes, sunt proxima primis
> sidera, suntque gradus atque omnia iusta priorum:
> maximus est populus summo qui culmine fertur;
> cui si pro numero vires natura dedisset,
> ipse suas aether flammas sufferre nequiret,
> totus et accenso mundus flagraret Olympo.

> There are stars similar to princes, and those stars that are closest to these leading figures, and there are all the grades and formalities of the more eminent men. Greatest in number is the populace which is carried around the highest dome of heaven: if nature had given them power to match their number, the ether itself would not now be able to withstand its own flames and the entire universe would now be burning with Olympus ablaze.

Ostensibly a comment about nature's conscious design for avoiding cosmic conflagration, Manilius' anthropomorphic imagery encourages us to consider more earthly scenarios, in which the masses might overthrow the Emperor if only they had power to match their number (5.743 *pro numero vires*).[97] Reassuringly, however, cosmic destruction is only revealed to us within an unreal conditional clause. Nature has not distributed power equally to all stars, and so all is well in the universe. In similar vein, Manilius can justly claim to be averting major imperial upheaval by withholding from the masses the authoritative power of horoscopic astrology.[98]

SUMMARISING THE STUDENT'S EXPERIENCE IN MANILIUS

After such a lengthy journey through the five books of Manilius' poem, it would be appropriate at this point to offer a summary of my

[97] Although they are referred to as *populus* here, Manilius uses more disparaging language of the 'common stars' at 1.470–1 (*omne stellarum vulgus*). In view of the popular connection between the Emperor and Jupiter, it is not difficult to read *accenso . . . Olympo* as a veiled threat to the principate.

[98] It should be implicit from my reading here that I do not consider there to have been additional books of the poem; for further discussion, see e.g. Landolfi (1991) 256 n. 45; Abry (2007b).

position. Reading Manilius' poem, I maintain, is not a didactic transaction from which the student only feels short-changed once he has finished reading the extant text and has realised that important parts of the instruction are missing. This in itself would not make Manilius' project particularly distinct from other ancient didactic poems, a genre in which instruction is never exhaustive.[99] What is distinctive, however, is the intratextual relationship between teacher and student, which unfolds as a progressive drama from a position of enthusiasm on the part of both parties in Book 1 to a level of confrontation in Book 4 when the student offers direct expressions of his despondency. This sense of student disillusionment, moreover, is not reconciled in Book 4, and the sort of undermining of instruction that irritated the student in Book 4 re-emerges in another guise in Book 5.

We have now to ask how such a student-centred interpretation of the poem fits with existing, largely teacher-centred scholarly viewpoints. Neuburg, while acknowledging pedagogical difficulties in Manilius' approach, maintains a positive learning outcome for the student: the teacher is seeking to inculcate an almost 'religious' feeling of awe and wonder in his students, bidding them to worship the absolute power of the stars rather than question any inconsistency between teaching and determinism (1993: 280–1). Volk reaches a similar conclusion, suggesting that Manilius is seeking not to teach astrology, but rather to create 'an alluring vision of astrology as a divine science, a vision that his readers can enjoy—or feel elated by—at the same time as they appreciate the poet's verse' (Volk 2009: 181); the didactic teacher, Volk maintains, is a dilettante in a range of philosophical approaches and rhetorical poses who draws arguments at will from different sources with little concern for authorial consistency.[100]

These views are indeed interesting, but they can only represent one side of the argument. Irrespective of the teacher's intentions, the viewpoint of the vociferous internal student demands attention, and it is clear from the 'outbursts' at 4.387–9 and 4.869–72 that the student has not succumbed to any acquiescent and reverent worship of the stars, nor does he recognise or appreciate the polyphony of his teacher. Such despondent feedback from the student—whom, as I have said, I am prepared to map closely on to the contemporary Roman

[99] See esp. Hunter (1995).

[100] See Volk (2002) 208; Volk (2009) *passim*; Volk (2011b). For a similar (though independent) view of Aratus as an author who presents in his poem competing stellar ideas; see Pendergraft (1990).

reader—legitimises a student-centred reading of the poem as a *failed* lesson: the primary Reader was indeed expecting this to be a lesson, and is frustrated to discover that it is not delivered.

In the light of the presence of the frustrated student, I must disagree with the very soft didacticism that Volk attributes to Manilius' poem when she compares *Astronomica* to modern coffee-table publications:

> It thus seems best to me to view didactic poems as a high-end ancient version of our coffee-table books . . . I suggest that didactic poems function in a similar manner: they are appreciated for evoking (not actually teaching) an interesting subject matter in an aesthetic way . . . We can thus conclude (and we may have suspected this all along) that Manilius' poem—despite all its 'sums in verse' and discussions of astrological minutiae—is not really about how to do astrology, but about the idea of astrology itself. (Volk 2009: 181)

> We have seen above . . . that . . . the *Astronomica* functions like a coffee-table book, which the reader picks up to delight in the idea of astrology without bothering overly much about its practical details. (Volk 2009: 262)

Application of such a modern analogy may well be appropriate for some of the more 'encyclopaedic' and technical prose works of the ancient world—such as Vitruvius' *De Architectura*, Strabo's geographical treatise and the *Natural History* of Pliny the Elder—which concentrate on setting out a subject rather than creating an instructional dialogue with a specified student or set of students. But didactic poetry is something different, as Hunter (1995) captures succinctly in his distinction between Aratus' didactic output and his astronomical prose sources:

> It is the choice of the didactic, poetic mode which precisely advertises the importance of the audience's role. Aratus' poem, unlike the work of Eudoxus or Hipparchus, is not merely about the universe, but is also universal in the sense that it presents itself as available to all, farmers, sailors, literary scholars.

Hunter's distinction is equally useful for our study of Manilius: one should not downplay the didactic intent of the teacher and the experience (in this case, frustration) of the specified learner(s). One may find uncomfortable the gulf between purported didactic intent and learning reality in our poem, but this is no reason for redefining the rules of the game, especially when one considers Manilius against a backdrop of texts which play out a didacticism of obscurity.

MANILIUS IN CONTEXT (1): THE DIDACTICISM OF OBSCURITY AND COMPETING READERSHIP

First of all, staying within the realm of astrology, one might say that Manilius is simply the first extant, and certainly the most elaborate, exponent of a teaching of astrology that is riddled with confusion and contradiction. As mentioned in the Introduction, Barton has observed similar tensions in the astrological 'handbooks' of the 1st to 3rd centuries AD: presentation of incompatible systems of astrological calculation, information overload, deferral of meaning and contradictory instruction, for example, are not the preserve of Manilius' astrological teaching.[101] In interpreting these complexities, Barton draws attention to the ways in which the astrologer-author fashions himself as a figure of unreachable knowledge for the novice student-reader.[102] Can we rule out the possibility that Manilius is motivated by a similar desire, to carve out for himself a position of power in the new imperial world of experts? We cannot rule this out, but I think the deference shown to the Emperor (esp. 1.7–10), along with Manilius' pride in poetic innovation rather than technical mastery (e.g. 3.1–35), points instead to a poet who fashions himself as a compliant imperial agent, intent on producing a creative poetic enterprise that plots its own way through the levels of acceptable stellar disclosure in the early empire.

Nor should the poetic medium of didactic prove to be an obstacle to such a reading. On the contrary, by the time the Romans had turned their hand to didactic poetry, the genre had already developed into a mode of discourse which was far more complex than is suggested by its surface façade of providing instruction to the student(s) in a poetic form. Rather, it had developed into a playful medium in which the author could create disjuncture between message (content) and medium (form and style); speak with various differing voices, providing 'polyphony' if not open contradiction; and speak to the interests of differing audiences.[103]

[101] See Barton (1994a) 71–94.

[102] See Barton (1994a) 82–5. For imperial authors of technical 'manuals' who fashion themselves as unreachable authority figures through complications and omissions in their 'instruction', see also the forceful reading of Frontinus offered by König (2007) esp. 191–7; for similar readings of Vitruvius, see also Chapter 6, n. 12.

[103] For various useful formulations of these dynamics in didactic literature, see Effe (1977) 26–39; Heath (1985) 253–5; Toohey (1996); Volk (2002) 44–68. See esp. Toohey (1996) for the developing 'playfulness' of the genre.

All three elements of play are in evidence in Manilius. The disjuncture between instruction and medium is most obviously felt in Book 3, where complex mathematical calculations are confined to hexameter and obscured behind poetic periphrasis.[104] Furthermore, as I have suggested above, Manilius' didactic voice is by no means uniform: not only does he include competing systems of astrological thought, but he suggests that astrological knowledge is both inclusive and very exclusive; moreover, in terms of style, I have argued that Manilius' surface endorsement of the student's ability to grasp the material is matched with more subtle but sustained attempts to undermine the student's current knowledge and morale. But it is the third element of play—speaking to the interests of different audiences—which is particularly important to an understanding of what Manilius is doing in his poem, and it is a tactic that he may well have picked up from his immediate predecessor, Ovid, in his erotodidactic works.

In general terms, one can readily detect the influence of Ovid's *Ars Amatoria* and *Remedia Amoris* on Manilius, since (on my reading) both authors set out to teach the unlearnable: how to fall in love, how to manage a relationship, how to rid oneself of love, and how to cast a horoscope are all 'arts' which are, for one reason or another, resistant to static instruction. But *Ars Amatoria* 3 in particular may have proven instructive for Manilius.

After two books instructing men on how to find, approach and keep hold of a woman, Ovid turns his attention in *Ars* 3 ostensibly to the amatory instruction of women (*Ars* 3.1–6):[105]

> arma dedi Danais in Amazonas; arma supersunt
>   quae tibi dem et turmae, Penthesilea, tuae.
> ite in bella pares; vincant, quibus alma Dione
>   faverit et toto qui volat orbe puer.
> non erat armatis aequum concurrere nudas;
>   sic etiam vobis vincere turpe, viri.

> I have given arms to the Danai against the Amazons; there remain the arms which I should give to you, Penthesilea, and your squadron. Go into battle on equal terms: let them conquer, those whom kindly Dione favours and the boy who flies over the entire world. It was not fair for armed men to fight against defenceless

[104] Manilius specifically draws attention to the challenge of fitting technical instruction into (hexameter) verse at 1.20–4, 3.31–42, 4.431–42.

[105] The text of Ovid is taken from Kenney (1961).

female opponents: in this way would it be disgraceful for you, o men, to conquer.

At first glance, this appears to be a (rare) victory for sexual equality in the ancient Roman world. But a closer look at the instruction offered throughout *Ars* 3 reveals otherwise.[106] Recent scholarship has become particularly attuned to the fact that the female addressees of *Ars* 3 are not really offered independent instruction at all. They are instructed to choose appropriate hairstyle (3.135–68) and clothing (3.169–92); to train themselves in the concealing of defects (3.193–290); and to enhance their appeal by adopting pleasing speech and gait and developing musical, poetic and dancing abilities (3.291–380). In effect, the instructor is grooming women to look attractive and take on a *passive* role in the relationship such that they become 'sophisticated accomplices' in the male game of love as expounded in *Ars* 1 and 2.[107]

Such male-oriented instruction in *Ars* 3 would perhaps naturally imply the presence of a male audience for the book. As we have seen, the direct address to men at the beginning of the book (3.6 *viri*) testifies to their continued presence from the previous two books. Nor should we simply assume that this audience leaves as soon as the instructor starts instructing women. Scholars have noticed how in certain transitional passages the *praeceptor* continues to court the attention of men:

nunc quoque nescirent! sed me Cytherea docere
iussit . . . (*Ars* 3.43-4)

quis credat? discunt etiam ridere puellae. (*Ars* 3.281)

parva monere pudet, talorum dicere iactus
ut sciat et vires, tessera missa, tuas. (*Ars* 3.353-4)

quo feror insanus? quid aperto pectore in hostem
mittor et indicio prodor ab ipse meo? (*Ars* 3.667-8)

I wish that now also women did not know how to love! But Venus commanded me to teach them . . .

Who would believe it? Girls even learn how to laugh.

I am ashamed to advise on small matters, to speak about the throwing of dice, and how she should know your powers, o flung counter.

[106] See discussion with Green (2006) 11–14.
[107] See Myerowitz (1985) 79–86, 97–101, 123–8, 134–41; Downing (1990).

> To where am I being borne in my madness? Why am I sending myself against the enemy with my breast unguarded and betraying myself with my own judgement?

These four moments of authorial reflection are clearly not addressed to the female audience, as in all cases women are referred to in the third person. Rather, these are moments away from the primary female audience in which the teacher-poet can express his disbelief at female behaviour as well as his sense of shame and madness at instructing the sexual 'enemy'. These may be private reflections, but their chauvinistic nature points more readily to the continued presence of an eavesdropping male audience, whose underlying interests the instructor continues to promote.[108]

Following on from the didactic playfulness of Ovid's *Ars Amatoria*, I would suggest that Manilius has purposely constructed two very different addressees in the work, an imperial dedicatee (Augustus and/or Tiberius) and a student (Reader). The Reader, with whom the contemporary Roman reader might readily want to identify himself, is eager to acquire astrological knowledge for practical application. And yet, as we have seen, the teacher ultimately cheats this student of that lesson. In so doing, Manilius actually serves the interests of the other addressee, the imperial dedicatee: though he does not address him directly beyond 1.7, we may envisage him as sitting quietly in the background, appreciating Manilius' novel enterprise in celebrating the power of astrology while keeping its secrets safe.

MANILIUS IN CONTEXT (2): THE AUTHOR AND HIS AUGUSTAN CONTEMPORARIES

Ultimately, then, I see Manilius' project as a highly politicised venture in concert with the Emperor, aided and abetted by the playfully compliant genre of didactic. Previous readings of the poem have advocated a positive political thrust to the *Astronomica*. It has been argued, for example, that the poem pays homage to the Emperor, his building

[108] See Miller (1993) 238–41; Gibson (2003) 19–21.

programme and the boundlessness of his empire.[109] Moreover, some scholars have put together the incomplete nature of the extant poem with the growing political sensitivity surrounding astrology in the early years AD to suggest that Manilius thought it respectful to stop or shelve his project before it ran into political trouble.[110] My view builds on these ideas, but rather than speak in terms of a simple distinction between writing and shelving, or a removal of specific elements, my student-centred approach has enabled me to uncover what I see as a much more embedded strategy of misdirection on Manilius' part throughout the poem.

However we view the poem, it would be unwise to downplay its political angle.[111] It is true that Manilius does not draw specific attention to the importance of, less still the nervousness surrounding, astrology in imperial circles, in marked contrast to the astrological treatise of his successor, Firmicus Maternus (*c.* AD 334), which devotes an entire section to the social, moral and political responsibilities of the astrologer.[112] But political readings are not dependent on specific reference to political machinery.[113] The topic of astrology in itself, combined with the presence of an imperial addressee, cannot but push the poem into a political arena. Moreover, the recent example of Manilius' predecessor in didactic poetry, Ovid, would have been a salutary reminder of the dangers of assuming political immunity from controversial topics handled in poetry. In attempting to teach the art of love-making in *Ars Amatoria*, and conscious of the political sensitivity surrounding

[109] For positive political readings, see e.g. Abry (2011); Gale (2011); Kennedy (2011). This is not to deny that there are interesting negative political readings of the poem. I share the concerns of DeNardis (2003: esp. 162–217) that standard flatteries to the power and achievements of the Emperor are (uncomfortably) tempered by the ineluctable power of fate over mankind and the fortunes of leaders and their cities: this is a poem in which, at the very least, there is contention over how much power the Emperor really has.

[110] See e.g. Bouché-Leclercq (1899) 553–4; Garrod (1911) lxiv. Herbert-Brown (2002: 126) suggests that the absence of discussion about the planets might have been a specific measure to ensure the political safety of the project.

[111] As does, most recently, Volk (2009) 170–3, 181 n. 15.

[112] The diviner is advised to offer astrological advice in public and avoid private consultation (*Math.* 2.30.3, 10) and to shun any enquiries relating to the Emperor (*Math.* 2.30.4, 7). Both warnings respond specifically to the legislation on astrology that goes back to AD 11. Firmicus also responds to more recent, Diocletian legislation which stipulated that the life of the Emperor was beyond the remit of the stars' power (*Math.* 2.30.5). See further Cramer (1954) 279–81; Sogno (2005) 170–6.

[113] See Kennedy (1993) esp. 34–9.

marriage and adultery, for which Augustus had instituted legislation, Ovid makes regular efforts to de-politicise his poetic material by expressly excluding married women from his readership.[114] And yet this poem, despite its surface apolitical stance, proved to be a significant factor in his exile, charged as it was with inciting adultery.[115] Writing in the aftermath of the relegation of one of Augustan Rome's most famous poets, it is very difficult to imagine that Manilius could have proceeded without a real sense of the political waters into which he was wading. For this reason, the obscurity of his lesson and the discretion he adopts towards the topic are, I maintain, conscious politically informed choices.

Even though there is widespread acknowledgment of the practical uselessness of Manilius' text as an astrological handbook, it is nevertheless a popular conceit to go along with Manilius' own verdict that he is the first Roman writer on astrology. On the strength of this analysis, however, I feel that it would be more appropriate to refer to Manilius as a writer who indulges in an elaborate discourse of disclosure and discretion with regard to astrology. Viewed in this way, Manilius loses his status as pioneer. In fact, he is operating within a well-established Roman literary and intellectual context, in which writers seek to negotiate an appropriate political line between celebration of the power of astrology and avoidance of inappropriate knowledge disclosure.

The second part of my study seeks to contextualise Manilius' authorial strategies by taking a more detailed look at the political development of astrology at Rome in the 1st century BC, and the subsequent variety of responses generated from its contemporary commentators.

[114] Cf. *Ars* 1.31–4, 2.599–600, 3.57–8, 613–14; also *Rem.* 385–6. But there may be good reason not to take these disclaimers at face value; see Miller (1993) 234–6; Sharrock (1994a) 109–13; Green (2006) 13–14.

[115] This is according to Ovid himself: cf. e.g. *Tr.* 2.241–54, *Pont.* 3.3.49–58.

Part 2

The Rise of Roman Astrology and Caesar's Comet

3

The Rise of Astrology in Rome

Astrologers and astrological activity in Rome can be traced back to the late 3rd century BC.[1] The evidence is inevitably sporadic and haphazard, but tends to suggest that astrology gained popularity with the lower classes. Ennius appears to testify to the early presence of charlatan astrologers roaming the Circus at Rome,[2] and popular awareness of the role of the stars over mortal life may be further demonstrated by the presence and sentiments of the constellation Arcturus in the prologue to Plautus' *Rudens* (*c.* 211 BC). Reactions to astrology among the elite at this time are, by contrast, discernibly mixed. General expressions of contempt (cf. Enn. *Iphigenia* fr. 95 Jocelyn) may not have been universal, if astrological curiosity influenced Marcus Claudius Marcellus' act of bringing from Syracuse to Rome two of the celestial globes of Archimedes (212 BC), one of which was housed in the temple of Honour and Virtue.[3]

As the 2nd century BC progresses, the elite appear to become more actively concerned with astrology in their attempts to curtail its practice among the masses. In his agricultural work, dated *c.* 160 BC, Cato the Elder warns his bailiff against consulting charlatan astrologers (*RR* 1.5.4). More famously, and on a more public scale, in 139 BC the praetor Cn. Cornelius Hipsalus banned astrologers who made

[1] For earlier historical reconstructions of the rise of astrology in Rome, see Bouché-Leclercq (1899) 546–50; Cramer (1954) 44–80; Liebeschuetz (1979) 119–26; Barton (1994a) 33–40; Barton (1994c) 147–54.

[2] Ennius ap. Cic. *Div.* 1.132, although there is difficulty here in discerning which words are truly Ennian; see Nice (2001). For a recent reassessment of the 'street astrologer' in Rome, see Ripat (2011) esp. 123–8.

[3] Cf. Cic. *Rep.* 1.21–2; Ov. *Fast.* 6.277–80; Plu. *Marc.* 19.4–6.

money by cheating with their false predictions people with 'frivolous and foolish minds' (*levibus et ineptis ingeniis*, V. Max. 1.3.3).[4]

It is difficult to pinpoint the motivations behind elite interventions in the period. A traditional approach has been to maintain a sharp distinction between the elite and the masses in terms of astrological belief: while the masses are perceived as unsophisticated, trusting and gullible, the elite are seen as more intelligent and contemptuous towards a phoney foreign religious art. More recently, however, it has been suggested that these examples provide evidence not of elite scepticism towards astrology, but of elite nervousness that religious power—which in official Roman divination was controlled by and spoke to the interests of the elite—might be located among the masses. This might, then, be a question of fear of perceived shifts in power within Roman society rather than belief.[5]

Understanding early astrological activity at Rome in terms of power dynamics helps to explain the transition to the first use of astrology by leading Romans in the 1st century BC. The first attested case occurs in 87–86 BC, when the consul Octavius, who attempted to hold Rome against the siege of Marius and Cinna, allowed the city to be captured because of his trust in a personal horoscope which had forecast his safety (Plu. *Mar.* 42.4).[6] A little later, Sulla is supposed to have recorded in his Memoirs a favourable horoscope which had been cast for him (Plu. *Sulla* 37.1). Though it is advisable to exercise due caution towards these anecdotes—Plutarch is, of course, writing two hundred years after the events he describes—the situation becomes much clearer in the late Republic, when we can rely on the contemporary testimony of Cicero. Cicero informs us (albeit mockingly) that by the late Republic, it was customary for horoscopes to be cast for the most prominent

[4] See further Cramer (1951) 14–17. Though the event may be accurately recorded, the vehemence with which Valerius Maximus speaks about greedy astrologers and their clientele may reflect his own times; see Herbert-Brown (2002) 121.

[5] See North (1990) 58–60, who puts forward the view in his discussion of the Cato passage. In the Cato passage—the only early piece of evidence for which we also have the context for comments on astrology—it is clear that Cato's advice forms part of a section which impresses upon the bailiff his subordinate position relative to his master (1.5). Cato's anxieties would be shown to be justified sixty years later, when a farm bailiff named Athenion initiates a large slave revolt in Sicily through his use of astrology; cf. Diod. Sic. fr. 36.5.1.

[6] It is interesting to note that this event coincides with a return of Halley's comet in July–August 87 BC; cf. Cic. *ND* 2.14, Plin. *Nat.* 2.92; see further Barrett (1978) 93; Ramsey (2007) 179, 188–9. Could this chance occurrence have played a part in the political development of astrology at Rome?

Romans, Crassus, Pompey and Caesar (Cic. *Div.* 2.99). While these men merely attract horoscopes, the first elite Roman practitioner of astrology appears to be P. Nigidius Figulus. Though he had a brief political career, reaching the praetorship in 58 BC, Nigidius is most famous for devoting himself to intellectual pursuits particularly in the realm of divination.[7]

These scattered anecdotes can provide a narrative of astrological development at Rome in the late Republic.[8] But the *reasons* for the rapid rise in astrology in the 1st century BC continue to be the subject of scholarly debate, and it is prudent to acknowledge a combination of factors at play.

First, it was the breakdown in the traditional Republican political institutions which facilitated the rise in astrology. Politics and religion had always been intertwined: 'there is a process of mutual reflection between religious and political institutions because they are aspects of the same power relations'. (Barton 1994a: 36). At the height of the Republic, political and religious power had been diffused among many individuals: in fact, there were no religious 'individuals', as each member worked strictly as part of his priestly college. When this political system started to collapse in the late 2nd century BC to make way for the powerful Roman individual, there was a corresponding rise in forms of divination that were both more personal (i.e. ones which consulted the interests of the individual rather than the state) and more flexible (i.e. ones which were capable of providing more than a simple yes/no response to a ritual question). Haruspicy and astrology were ideally suited to catering to this religious need.[9] And so we find that the first diviners to be known by name are attached to the

[7] His works include two pieces on the oriental and Greek heavenly *sphaera*; for his life and divinatory works, see especially Cramer (1954) 63–5; Rawson (1985) 309–12. He is famously reputed to have made for the newly born Octavian (63 BC) an astrological prediction of his future domination (Suet. *Aug.* 94.5, Dio 45.1.3–5) and is given a public astrological role at Luc. 1.639–72, where he offers a detailed interpretation of the prodigies which accompany Caesar's domination of Italy in 49 BC; for the astrology of Nigidius in this passage, see Getty (1941) and (1960) and n. 18.

[8] In addition, it is worth noting the astrologically charged titles given to four of the mimes of Decimus Laberius, fl. 50's and 40's BC (*Aries*, *Taurus*, *Cancer* and *Virgo*). It is tempting to speculate that these dramas may have provided some sort of comic response to Rome's growing interest in astrology, for the amusement of both the upper and lower classes. Unfortunately, the surviving fragments offer no evidence either way; see Panayotakis (2010) 130–1, 157–8, 394.

[9] For scepticism concerning the ability of astrology to fulfil this political function, however, see Habinek (2007) 229–31.

first powerful Roman individuals: Gaius Gracchus is our first attested example, as he apparently had his own *haruspex* named Herrenius Siculus (V. Max. 9.12.6); Sulla would also later employ the services of the haruspex C. Postumius (Cic. *Div.* 1.72).[10]

Another important factor in the rise of astrology in the 1st century BC was the cultural, and more specifically philosophical, influence from Greece. During the period 170–140 BC, a series of Greek philosophers who held strong and divergent views towards astrology spent time among the elite at Rome.[11] If the scepticism towards astrology advocated by Carneades and Panaetius was to hold sway in the mid-to-late 2nd century BC, a notable change would occur with Posidonius, a philosopher with great enthusiasm for astrology, who arrived at Rome as an ambassador of Rhodes in 87–86 BC.[12] He was apparently well known to Pompey and Cicero,[13] and it is at least an interesting coincidence that Posidonius should visit Rome within the same year as our first attested use of astrology by an elite Roman (Octavius).[14] Greek cultural influence more generally brought stars into the mainstream. Of particular note is the influence of Alexandrian poetry among the Roman poets, and the special status afforded to Aratus' *Phaenomena*, a poem which was translated by many notable Romans from this period onwards, including Cicero, Germanicus and Ovid.[15]

By 44 BC, then, it would be fair to say that astrology and the stellar world more generally had become well established in the discourse

[10] See further North (1990) 65–71, who notes that individually named diviners occur during all phases of dynastic rule in Rome, both the mythical regal period and the late Republic onwards.

[11] The most important are Crates of Mallos, the Babylonian Diogenes, Carneades of Cyrene, Critolaus of Phaselis and Panaetius of Lindus. See especially Cramer (1954) 50–8; Morford (2002) 14–33.

[12] For Posidonius, see Cramer (1954) 61–3. It is difficult to know what to make of the potential role of the obscure Manilius Antiochus, who apparently came to Italy as a slave and was hailed by Pliny as an influential figure for astrology (Plin. *Nat.* 35.199).

[13] Both may have heard the great philosopher in Rhodes, and Cicero is referred to as a student of his; Plu. *Cic.* 4.4–5; *Pomp.* 42.5; also Cic. *Fat.* 5, where Cicero might be referring to Posidonius as his philosophical master. Posidonius is subsequently regularly referred to as a friend to both; cf. Cic. *Fin.* 1.6; *Tusc.* 2.61; *ND* 1.123; *Div.* 1.6; Strab. 11.491.

[14] Both Cramer (1954: 61–3) and Liebeschuetz (1979: 119–26) appear to regard Posidonius' influence as the most significant factor in the rise of astrology in Rome during this period; for a more cautious assessment, see Long (1982).

[15] On the popularity of Aratus among the Romans, see especially Chapter 8.

of the elite Roman intellectual.[16] There is also anecdotal evidence to suggest that astrology was playing (unspecified) parts in Roman politics from 87 BC. It is important, however, not to overstate this last point. Liebeschuetz (1979: 122) offers the valuable observation that, if one respects Cicero's contemporary testimony, one must also acknowledge that in his most socio-political writings—his speeches and letters—Cicero never uses the term *mathematicus*, the most neutral term for an astrologer, and only twice uses *astrologus*; the terms appear much more frequently, however, in his philosophical writings. This might suggest that in the period of Cicero's output (*c.* 81–43 BC), the astrologer played little or no part in the daily political affairs of the elite, for whom the issue of astrology was 'academic rather than practical' (Liebeschuetz 1979: 122).[17] Other ancient evidence—or rather its silence—bears out this impression. There is no direct evidence to suggest that leading figures in the late Republic included an astrologer as part of their personal entourage, nor that Rome made public use of astrologers in its own affairs.[18] Julius Caesar is a classic example. His enthusiasm and respect for scientific astronomy is well attested both in his astronomically informed reconfiguration of the Roman calendar in 47–46 BC—facilitated by the Alexandrian astronomer Sosigenes—and in the (simultaneous?) publication of his own astrometeorological treatise, *De Astris*.[19] And yet, in spite of Cicero's comment that a horoscope was cast for him (*Div.* 2.99), it is noteworthy that Caesar appears to have continued the practice of choosing personal diviners not from

[16] As well as the works of Nigidius Figulus and Cicero's extended discussions of astrology in *De Divinatione* and *De Fato* (for which see Chapter 4), cf. also Varro's (lost) chapter on astronomy/astrology in his *Disciplinae* (Cramer 1954: 66–7; Rawson 1985: 163–4).

[17] Even in the two instances of *astrologus* in Cicero's letters and speeches, there is nothing said about Roman astrology, as one refers to practice in Greek Sicily (*Ver.* 2.2.129) and the other (*Fam.* 6.6.7) amounts to a generic passing reference.

[18] Lucan's depiction of Nigidius Figulus offering a public astrological prediction in 49 BC (1.639–72) has intrigued scholars, but there is general consensus that Lucan is here engaging in significant poetic embellishment/fabrication; see Morford (1967) 63–4; Hannah (1996), esp. 185–6; Lewis (1998) argues for a contemporary Neronian resonance to the prophecy.

[19] For the Caesarian reform of the calendar, see Herbert-Brown (1994) 15–26; Domenicucci (1996) 85–8; Feeney (2007) 151–4; for the possible nature of the *De Astris*, the information for which was probably also supplied by Sosigenes, see Macr. *Sat.* 1.16.39; Cramer (1954) 74–7; Rawson (1985) 112, 165; Domenicucci (1996) 85–99; see also Chapter 8.

the astrologers but from the *haruspices* right up to his death.[20] Though one would have to wait a while for the first court astrologer (see Chapter 5), the credibility of astrology within a political context was about to take a decisive turn.

CAESAR'S COMET (JULY 44 BC) AND THE IMPERIAL AGE OF ASTROLOGY

In mid-July 44 BC, a chance occurrence—albeit brilliantly manipulated—would change the political landscape for astrology at Rome forever.[21] During the funeral games in honour of the recently deceased Julius Caesar, attached to the *Ludi Veneris Genetricis* (20–28 July),[22] a bright comet was visible at Rome over a period of seven nights and days.[23] Comets required religious explanation, and as interpretation of comets fell within the remit of both haruspices and astrologers, the event would prove to be a critical point in the struggle for supremacy between these two forms of divination. Both Cassius Dio (45.7.1–2) and Servius (on Verg. *Ecl.* 9.46) provide us with detail which might help us to recreate the spirit of divinatory competition on that day. Servius relates that a soothsayer named Vulcanius announced that the comet indicated the end of one age and the beginning of another. It is unlikely that Vulcanius' enigmatic interpretation would have been greeted as a positive omen, as comets were predominantly treated as negative predictions of (violent) changes in government and

[20] For Caesar's haruspex Spurrina, cf. Cic. *Div.* 1.119; Suet. *Iul.* 81; V. Max. 8.11.2. In this context, it is also worth noting that, in 43 BC, popular verses ridiculing the rise of Ventidius Bassus to consul call upon augurs and soothsayers, rather than astrologers, to explain this 'portent': *concurrite omnes augures, haruspices!* (ap. Gel. 15.4.3).

[21] In what follows, I am indebted to the fascinating interdisciplinary reassessment of the Julian comet by Ramsey and Licht (1997). See also Weinstock (1971) 370–84; Kronk (1999) 22–4.

[22] For the controversy surrounding the name and length of the festival in 44 BC, see Ramsey and Licht (1997) 41–57.

[23] Our main sources for the comet are Suet. *Iul.* 88; Dio 45.7.1–2; Plin. *Nat.* 2.93–4 (who purports to be making use of Augustus' Memoirs) and Servius on Verg. *Ecl.* 9.46. Details about the comet, then, come exclusively from later sources—the earliest being Augustus' Memoirs of *c.* 24–23 BC—and there is no contemporary Roman endorsement (most notably from Cicero), a situation which has invited questions about its veracity: might the comet be a later, Augustan fabrication for political advantage? We can never be sure, but both logical argument and external (Chinese) sources favour its veracity; see esp. Ramsey and Licht (1997) 61–74.

death to its leader(s).[24] But a different and more positive interpretation took hold of the popular imagination: the comet was interpreted as affirmation that the recently deceased Caesar had become a god.[25]

The possible reasons for the popularity of this new interpretation of a comet are discussed by Ramsey and Licht (1997: 135–53). In particular, one cannot ignore the context for the comet. These were after all the funeral games of Julius Caesar held during a festival to Venus, and it is argued that the star of Venus might also have been visible during at least part of the comet's course.[26] In the light of this fortuitous temporal and stellar context, it is not difficult to see how stories of Caesar's divination, aided by his mother Venus, might have arisen. The particular brilliance of this comet, visible even in daylight, might have lent it a divine glow and might have distinguished it further from its more baleful counterparts.[27] But such widespread consensus suggests overarching orchestration, and it is easy to believe Servius when he intimates that the politically astute Octavian was guiding the people towards this interpretation.[28] Consistent with the political mood, contemporary Augustan writers entertain various elaborate connections between this stellar phenomenon and Caesar's divinity, but never openly challenge Caesar's status as a god.[29]

Albeit a politically expedient move, the adoption of such an interpretation of the comet might also point to early astrological enthusiasm on the part of the future Emperor. After all, the translation of humans to the stars—and hence, more generally, the influence of earthly life on the stars—might be seen as a reciprocal feature of the

[24] For comets signifying violent change in leadership, cf. Luc. 1.526–9; Tac. *Ann.* 14.22, 15.47; Suet. *Claud.* 46, *Nero* 36 *stella crinita, quae summis potestatibus exitium portendere vulgo putatur*, *Vesp.* 23.4; Sil. 8.637 *regnorum eversor rubuit letale cometes*; see also Rawson (1978) 146; Ramsey (2007) 193–4. Moreover, the specifics of Vulcanius' prophecy—the coming of a tenth age—might have signalled the more worrying prospect of the end of the nation; see Ramsey and Licht (1997) 140–2.

[25] For contemporary intellectual enquiry as to whether the comet was a sign or a physical manifestation of Caesar's translation to the heavens, see Chapter 9.

[26] See Ramsey and Licht (1997) 138–9.

[27] See Ramsey and Licht (1997) 137–8; see also Gurval (1997) 44 n. 22.

[28] Cf. Serv. *Aen.* 6.790 *stella per diem apparuit, quam persuasione Augusti Caesaris esse populus credidit*; Serv. *Aen.* 8.681 *quod sidus Caesaris putatum est Augusto persuadente*; Serv. *Ecl.* 9.46 *ille eam esse confirmavit parentis sui . . . quam quidam ad inlustrandam gloriam Caesaris iuvenis pertinere existabant, ipse animam patris sui esse voluit.*

[29] See further Chapter 9.

more conventional symbiotic relationship between stellar and earthly worlds.[30] There is anecdotal evidence to suggest that earlier in that same year, before he left Apollonia (March 44 BC), Octavian made use of the services of an astrologer named Theogenes to cast his horoscope, and that he was pleased with the result (Suet. *Aug.* 94.12).[31] Two sources confirm that Octavian subsequently published his horoscope, but they make it difficult to determine exactly when this took place. Cassius Dio seems to suggest that the horoscope was made public in the same year as the edict proscribing astrological consultations about death, namely AD 11 (56.25.5):[32]

> τότε δ' οὖν ταῦτά τε οὕτως ἐπράχθη, καὶ τοῖς μάντεσιν ἀπηγορεύθη μήτε κατὰ μόνας τινὶ μήτε περὶ θανάτου, μηδ' ἂν ἄλλοι συμπαρῶσίν οἱ, χρᾶν καίτοι οὕτως οὐδὲν τῷ Αὐγούστῳ τῶν καθ' ἑαυτὸν ἔμελεν ὥστε ἐκ προγραφῆς πᾶσι τὴν τῶν ἀστέρων διάταξιν, ὑφ' ὧν ἐγεγέννητο, φανερῶσαι.
>
> Besides these events at the time, seers were forbidden to prophesy to anyone either alone or, on the subject of death, even if others should be present. Yet Augustus was so wholly indifferent where he was himself concerned that he published by edict for all to know the arrangement of the stars under which he had been born. Still Augustus did issue the prohibition in question.

By contrast, in the same section in which he dealt with the meeting with Theogenes, Suetonius suggests an earlier date for public dissemination of the horoscope (*Aug.* 94.12):[33]

> tantam **mox** fiduciam fati Augustus habuit, ut thema suum vulgaverit nummumque argenteum nota sideris Capricorni, quo natus est, percusserit.
>
> Augustus **in due course** had so much faith in his fate that he made public his own horoscope[34] and struck a silver coin with the mark of the constellation of Capricorn under which he was born.

Notwithstanding the semantic range for *mox*, which can be used to indicate causal as well as temporal connection,[35] it is highly unlikely

[30] See Cramer (1954: 78), who suggests that catasterism belongs to a branch of 'religious' rather than 'scientific' astrology.

[31] For the scientific credentials of Theogenes, see Kennedy (2011) 169–70.

[32] The text of Dio is taken from Boissevain (1955–69); the translation is influenced by Swan (2004) 280–1.

[33] The text of Suetonius is taken from Ihm (1964).

[34] For *thema* as 'horoscope', cf. Sidon. *Ep.* 8.11.9.

[35] See Rose (1927), esp. 63–4 for the varied use of *mox* in Suetonius.

that Suetonius is alluding to the date of AD 11 that we find in Cassius Dio. Within the temporal parameters of a biography, *mox* would seem a most inappropriate means for Suetonius to connect events separated by nearly sixty years of an individual's life. The reference to the coin which Augustus had struck backs up this impression. The first undisputed official numismatic connection between Augustus and Capricorn can be dated to the eastern mint of 28 BC, and there may have been coins issued to this effect as early as 41/40 BC.[36] Indeed, the order of thought in Suetonius is important: placing the detail about the horoscope before the coin implies that the horoscope was made public before the minting of the coin, an order which seems perfectly natural if Augustus wanted people to recognise the significance behind the coin's symbolism.

So the sources would appear to present two distinct options for the dating of the publication of the horoscope, although many scholars have been guilty of disregarding Suetonius (or at least his use of *mox* and the ordering of his thoughts) and plumping for Dio's date of AD 11, where the publication might have been designed to allay fears of the Emperor's imminent death.[37] It is important to be aware, however, that Dio does not specifically give a date for the crucial event in question. As Potter (1994: 147 n. 4) duly remarks, Dio mentions the horoscope primarily as an example of Augustus' lack of consistency in astrological matters; he does not strictly tell us that this event was concurrent with the official ban on astrological readings. I am inclined to agree with those who follow Suetonius' lead in positing a much earlier date for the publication of the horoscope, at least in general terms.[38] Some would like to place the horoscope's publication very early, where it might form the first and certainly the most overt example of the young prince's emerging personal astrological

[36] See Chapter 5.

[37] For this viewpoint, see e.g. Cramer (1954) 83; Liebeschuetz (1979) 123; Abry (1988) 103 n. 1; Barton (1994c) 153; Barton (1995) 34–5; Herbert-Brown (2002) 117. MacMullen (1967: 131–2) is one of the few earlier scholars to remain cautious on the issue.

[38] Nor is it the case that to agree with Suetonius is to disregard Cassius Dio. No previous scholar has, to my knowledge, recognised that the two sources can in fact be reconciled if we interpret Augustus' actions in AD 11 as a *reaffirmation* of his positive horoscope at a time of uncertainty about the Emperor's future, or as an *official* confirmation of something that had been made public earlier in less formal terms: note that Dio speaks of an edict (ἐκ προγραφῆς), whereas Suetonius may be implying in his use of *vulgaverit* a more general, earlier divulgence of the contents of the horoscope.

discourse.[39] It is worth noting, however, that there is nothing in Suetonius to rule out the possibility that the publication took place very soon after his meeting with Theogenes, perhaps even in the same year.

In terms of both personal enthusiasm and political expediency, therefore, it would have been fitting for Octavian to turn to an astrological response to the comet: in contrast to the haruspices, astrologers often found, as here, positive interpretations in these phenomena.[40] The Julian comet, then, can be seen as the pivotal event in the official endorsement of astrology in Roman politics, embraced simultaneously by both the leading figure in Rome and the people for perhaps the first time in Rome's history. Following on from this, Caesar's consecration and place among the gods of the state were formally decreed by the Senate on 1 January 42 BC.[41] If astrology and its exponents had been background players in the politics of the late Republic, they would henceforth be firmly in the forefront of Roman political discourse under the empire.

[39] Wray (2001) believes it might have been published 'long before Augustus came to sole power'; Potter (1994: 147) believes that the publication may have coincided with the principate in 27 BC, and Ramsey and Licht (1997: 148 n. 42) likewise prefer a date in the early 20's BC.

[40] See Bouché-Leclercq (1899) 357–62; Ramsey and Licht (1997) 145–7.

[41] For reconstruction of the event, see Weinstock (1971) 386–98.

4

The Beginning and End of the Late Republican Astrological Debate

The Politicised Philosophical Posturings of Cicero

In Chapter 3, it was argued that for the elite Roman of the late Republic, astrology was a private academic rather than public political issue. It is entirely appropriate, therefore, that the first extant discussion of astrology by a Roman should occur in a philosophical work, namely Cicero's *De Divinatione* (2.87–99). The topic then resurfaces briefly in Cicero's subsequent philosophical work, *De Fato* (7–8), but with a noticeably different emphasis. In what follows, I will discuss these two sections in detail, paying particular attention to the difference in approach to astrology shown in each work and the possible political reasons behind this.

ASTROLOGY IN *DE DIVINATIONE*: CICERO ON THE OFFENSIVE

De Divinatione forms part of Cicero's trilogy of philosophical works which deal with the relationship between gods and human affairs: it forms a planned sequel to *De Natura Deorum*, in which a more specific focus on divination is deferred for another time (3.19), and precedes the third member, the now fragmentary *De Fato*.

The dating of *De Divinatione* is not the subject of significant dispute. On the one hand, there are some direct references to the (recent) death of Julius Caesar in the work; on the other, there are parts where either a reference to Caesar's death is curiously absent, or else the sentiments might seem more appropriate if Caesar were still alive.

Consequently, there is general consensus that the work was written over a short period of time: started in late 45 or early 44 BC, it was essentially complete by the time of Caesar's assassination in March 44 BC; certain parts were revised or added in the few months after Caesar's death—most notably the Preface to Book 2 (1–7)—and the whole work was probably published before May 44 BC.[1] We can be reasonably confident that it was completed before the appearance of the comet that shone during Caesar's funeral games.[2]

A treatise in two books, the work takes the form of a dialogue between two individuals, Marcus and his brother Quintus, and has a discernible, if varied, structure.[3] In Book 1, Quintus puts forward the view in support of the power of divination, and as such presents himself as the upholder of a position that has both general consensus and ancient authority behind it (cf. e.g. 1.1, 2, 11). Skirting round the issue of exactly *how* divination works, Quintus concentrates on *what* happens (1.12, 16, 35, 109), and spends most of his speech presenting the listener with a series of examples and anecdotes, taken from both myth and history, in an attempt to prove the efficacy of divination throughout many centuries.[4] Quintus, it would seem, is seeking to convince by a preponderance of evidence (cf. 1.60, 71). In Book 2, by contrast, a character called Marcus appeals to reason, and sets about dismissing the examples advanced by Quintus. In a more systematic order than that adopted by his brother, Marcus dismisses the entire range of divinatory practices, both artificial and natural, including astrology.

This brief summary glosses over a major scholarly dispute in the text, namely the identity of the 'Marcus' who speaks in Book 2 or, more precisely, the extent to which he can be mapped onto the historical author Cicero.[5] As I will be arguing later for a shared consciousness between the character of Marcus in *De Divinatione* and the narrator-figure in

[1] The case for this dating of the work was originally made by Durand (1903). Durand's thesis was subsequently accepted by Pease (1920–3: 13–15), and is still the most plausible reconstruction; see now Wardle (2006) 37–43.

[2] In particular, the negative connotations associated with comets found at *Div.* 1.17 are not tempered anywhere else in the work, which might be felt to be too gratuitous if the work arose in the wake of Caesar's comet.

[3] See esp. Schofield (1986) 51–5; Krostenko (2000) 370–3; Wardle (2006) 21–3.

[4] Indeed, Quintus' case seems to be 'palpably designed to appeal to the Roman reader and with comparatively little exposition or criticism of Greek philosophical positions' (Schofield 1986: 50). That is not to say, however, that Quintus' argument is not clever in its underlying rhetoric; see Krostenko (2000) 370–3.

[5] In the ensuing discussion, I shall use 'Marcus' to refer to the character constructed in the text, and 'Cicero' to refer to the historical author.

De Fato, it is necessary first to justify the overlap of Marcus with the historical author in the current work.

Older scholarship tends not to observe any distinction between internal character and historical author in *De Divinatione*: the arguments in Book 2 are taken simply to be the views of the real Cicero in 44 BC. Such an interpretation, however, requires justification, since an anti-divinatory stance might appear, at first glance, to sit uncomfortably with the conservative Cicero, who elsewhere champions divination and had been an augur in a professional capacity since 53 BC.[6] Various explanations are offered in an effort to account for this apparent inconsistency. Some seek to draw a contrast between public and private opinion, namely, between Cicero's public persona—which recognises the utilitarian value of divination in ensuring social cohesion and maintaining elite power—and a private, philosophical pose that rejects notions of deities intervening in mortal affairs.[7] Others stress the importance of the late date of *De Divinatione*, and argue along various lines that Cicero has changed his viewpoint since the earlier works. He may, perhaps, have altered his philosophical stance during the intervening period.[8] More interestingly, it is argued that Cicero may have been reacting to the changing socio-political climate. The abuse of religion under Julius Caesar, in particular the gradual passing of divinatory power from the state-owned priestly colleges to the powerful political individual, might have affected Cicero's view on divination: as it was no longer designed for the benefit of the state and social cohesion, it could be (arguably needed to be) duly attacked.[9]

[6] The most telling comparison can be made between *Div.* 2.70–4, an attack on augury, and *Leg.* 2.31–33, where Cicero expresses a strong belief in the practice. For other apparent direct contradictions, cf. e.g. *Div.* 2.28–69 with the author's positive comments about haruspicy in *Har. Resp.* 18–20; see further Wardle (2006) 5–8.

[7] For the utilitarian versus philosophical viewpoints, see Pavis d'Escurac (1981) 34 and Linderski (1982) 22–3. For the distinctly private setting chosen for the *De Divinatione*—a leisurely conversation between brothers at Cicero's Tusculan villa—cf. *Div.* 1.8, 10–11; see further Momigliano (1984) 208–9.

[8] Schofield (1986: 47–8) suggests that Cicero may have switched allegiance to the new Academy because of its greater potential for intellectual debate. For the view that Cicero changed philosophical affiliation twice during his life, see Glucker (1988).

[9] See Linderski (1982: 34–8), who refers to Cicero's stance as 'philosophically sound and politically prudent' (38). Momigliano (1984) comes to similar conclusions independently, but gives the political reading added force by suggesting that Cicero was also reacting to Varro's apparent acquiescence towards Caesar's religious manipulation: 'What matters is that Cicero became more skeptical when his contemporaries became more credulous or at least more sanctimonious' (205). A politically oriented approach to the text has also found favour more recently with Krostenko (2000), esp. 383–5.

A major re-examination of the traditional position has been offered by Beard (1986), who stresses that more weight should be given to the literary character of the work. Beard acknowledges an apparent inconsistency of thought on religious matters across Cicero's works but argues that this should impress upon us the futility of connecting the character 'Marcus' with the author Cicero: Marcus should be regarded as just another creation from the author (Beard 1986: 35). Beard also draws attention to the structure of the work, which might be seen to present an even-handed debate, in that each side of the argument is handled in a single book and is advocated by protagonists of comparable social status. This viewpoint may be strengthened by consulting the finale to the work (2.150), which can be interpreted as providing no guided conclusion to the issues raised in the debate: as is appropriate to the philosopher of the New Academy, objections are raised but the final decision is left for the audience (Beard 1986: 34–5). For Beard, then, *De Divinatione* is not a treatise stacked heavily in favour of its author's own sceptical views on divination in Book 2. Rather, the open-ended nature of the discussion is to be read as a celebration of Roman religious debate, perhaps pioneered by Cicero, as he is the first surviving author to apply the arguments of Greek philosophy to Roman cultic practice and thought (Beard 1986: 38). The work bears witness, then, to an important stage in the cultural development of the late Republic, as Rome develops its 'discourse on theology' (Beard 1986: 46).

Although it is an important study in many respects, Beard's central view that Marcus cannot be identified with Cicero pushes too far a cautionary stance towards literary persona. Granted that the identification is not obligatory, it seems an even odder choice not to entertain a connection, in the light of the specific literary choices that Cicero has made for the work. No previous extant treatise on divination takes the dialogue form, and it must have been a conscious literary choice on the part of Cicero to opt for a contemporary setting for the discussion and to include 'himself' as a character.[10] Moreover, *De Divinatione* actually works very hard to counter any sense of distance

[10] Cicero could hardly have been oblivious to the implications of offering sentiments through a character of 'himself'. In a letter to his brother in 54 BC, Cicero reports that his friend Sallustius, commenting on a reading of a preliminary draft of *Republic*, recommended that Cicero voice sentiments in his own person precisely so as to give it more authority; cf. Cic. *Q. fr.* 3.5.1–2. See further Wardle (2006) 11–12.

between internal character and historical author. In Book 1, Quintus refers to Marcus as an augur (1.25), and offers direct quotations from or allusions to Marcus' works, which include a translation of Aratus (1.13), a poem on his consulship (1.17–22), the *De Natura Deorum* (1.33) and a poem on Marius (1.106). The links between Marcus and Cicero are all too clear, since Marcus the character, Cicero the augur and Cicero the writer all appear to share the same persona in *De Divinatione*.[11] It is no surprise, then, to find Quintus noting the potential for clash between the public role and anti-divinatory stance of Marcus/Cicero (1.22):[12]

> tu igitur animum poteris inducere contra ea, quae a me disputantur de divinatione, dicere, qui et gesseris ea quae gessisti, et ea quae pronuntiavi accuratissume scripseris?

> Could you, therefore, be persuaded to speak against my arguments on divination, you who have done the things you have done and have written with utmost care about the things which I have discussed?

Caution should also be advocated in reading the conclusion to the work as necessarily open-ended.[13] In particular, as Wardle (2006: 15–16) notes, it was the role of the Academic philosopher simply to expose the weaknesses in a philosophical case, as Marcus does with Quintus' pro-divination argument. That Quintus is allowed to hold his own view at the end (2.150) is, therefore, a characteristic of the Academic school, which resists imposing a viewpoint on others: it does not, however, constitute an admission that all views are equally valid.

If Beard's study has not encouraged scholars to reject the identification of Marcus with Cicero, it has at least challenged them to give more serious consideration to the literary dimension of the text in justifying this identification. An attractive, blended approach has been adopted by Schofield (1986: 61–3), for whom *De Divinatione* is a multi-layered text. Though Schofield is inclined to favour the view that Marcus is to be seen chiefly as a philosophical player in a debate, he acknowledges that there is significant force to the assumption that at least some readers would have taken Marcus' views as representative of the author's

[11] See also Schofield (1986) 56.
[12] The text of Cicero is taken from Ax (1977).
[13] See Wardle (2006) 14–17.

real views (and that Cicero might have intended this). The reader is therefore left free to decide: is the work simply a celebration of newly founded Roman religious debate, or does it offer direct insight into the real, anti-divination views of Cicero?[14]

My own viewpoint, while in essence falling between several scholarly positions, leans towards the socio-political approach advocated by Linderski (1982) and Momigliano (1984). As I see it, the literary framework of the *De Divinatione* opens the text up to be viewed as a multi-faceted, socio-political document of its time: in the foreground, and specifically in Book 2, we may hear the views of its author, as he tailors his beliefs to the fast-changing, political relevance of divination; in the background, at a more abstract level, the work operates in a spirit of intellectual debate, recently facilitated by the application of Greek philosophy to Roman issues, which celebrates Roman religion as an important new topic for academic discourse.

But what does all this mean specifically for our interpretation of Marcus' discussion of astrology in *Div.* 2.87–99, a topic which is typically touched upon lightly in scholarly discussions of the work's broader issues? If we are justified in taking Marcus as representative of Cicero's opinions, the text may be viewed as a fruitful source to assess the socio-political status of astrology (just) before July 44 BC.

First, it is important to note an unusual quality about the discussion of astrology at 2.87–99. Viewed from one angle, the discussion is quite out of place in that, unlike the other examples of divination treated in Book 2, Marcus is not responding directly to anything that Quintus has said in Book 1, for whom astrology had warranted only cursory glances.[15] The treatment of astrology in Book 2 is motivated not by any specific intra-textual prompting from Quintus, but rather from the growing extra-textual prominence that astrology was assuming in elite Roman intellectual circles.

That said, it is equally important to note the consistent manner in which scientific astrology is treated as a distinctly foreign divinatory practice in the work, clearly demarcated from forms of Roman

[14] These are by no means the only options available. For example, some scholarship argues that Cicero has designed Marcus' speech, or indeed both speeches, to be rhetorically flawed; see Denyer (1985); Krostenko (2000).

[15] The only point of contact on astrology between Quintus and Marcus occurs at 2.97, where Marcus recalls Quintus' passing comment about the Babylonians' record keeping and testing of horoscopes (1.36).

divination.[16] Natal astrology is the first specific form of divination to be mentioned (1.2), where it is acknowledged as the oldest form and is firmly assigned to the Assyrians—by virtue of their flat terrain which allowed unobstructed views of the heavens—the Chaldeans and the Egyptians.[17] This is clearly set apart from practices in Italy (1.3), where augury takes the place of astrology as the oldest form of divination; Rome will subsequently expand its range of divinatory practices to include haruspicy, Sibylline verses and dreams, but no mention is made of astrology (1.4). The distinctly foreign nature of astrology is maintained in the work as a whole. In the course of Quintus' speech (1.91–3), a slightly different configuration of foreigners—Chaldeans, Egyptians and Babylonians (cf. 1.36)—emerges as practitioners of astrology, and further stress is laid on the differing terrain of Italy and the East to explain why astrology is not native to Italy.

It is noteworthy that Quintus should assign a foreign quality to astrology, considering his general tendency to blur distinctions between Roman and foreign divination.[18] Consistent with this strategy, Quintus adduces no anecdote for the use of astrology by any famous Roman individual. This must surely have been a conscious decision on Cicero's part, in the light of the examples he might have used. It is particularly telling that Quintus does not bolster his case by making reference to Nigidius Figulus, the most famous contemporary intellectual and practitioner of astrology.[19]

Allied to this strategy is a clear demarcation between a respectable, scientific mode of astrology, associated particularly with the East, and a more vulgar, unsophisticated type practised in the Circus at Rome.

[16] It is interesting to note that verses from Cicero's own *De Consulatu suo* (1.17–18) describe the Muse Urania extolling the virtue of astrological observation in the Roman context of Cicero's consulship in 63 BC, but the fragment is too de-contextualised to enable us to make any links with Cicero's own views at the time.

[17] Cicero's phrase *ab ultumis* (1.2) could operate in both a temporal and spatial sense, denoting a practice which is both very old and geographically distanced from Rome.

[18] See Krostenko (2000) 361–5.

[19] Nigidius Figulus had been exiled by Caesar and had died in 46–45 BC, before he could be pardoned. As far as we can tell, however, it was his conservative, senatorial politics which proved to be his undoing with Caesar rather than his intellectual/religious activities: cf. esp. Cic. *Fam.* 4.13.4, where Cicero suggests to Figulus that contemplation of his intellectual achievements may prove a solace to him in exile. He could, therefore, have been safely cited by Quintus as an example of Roman astrology if Cicero had so wanted.

De Divinatione only classes as divination astrology which is scientific (dependent on complex calculations) and strongly deterministic (in the sense that the stars are believed to govern *all* human action in advance).[20] Excluded from discussion is the astrology peddled for gain by superstitious amateurs around the Circus, for whom Quintus makes use of Ennius to register his contempt (1.132):

> nunc illa testabor, non me sortilegos neque eos qui quaestus causa hariolentur, ne psychomantia quidem, quibus Appius amicus tuus uti solebat, agnoscere; non habeo denique nauci Marsum augurem, non vicanos haruspices, **non de circo astrologos**, non Isiacos coniectores, non interpretes somniorum; non enim sunt hi aut scientia aut arte divini sed
>
> 'superstitiosi vates impudentesque harioli
> aut inertes aut insani aut quibus egestas imperat,
> qui sibi semitam non sapiunt alteri monstrant viam;
> quibus divitias pollicentur, ab iis drachumam ipsi petunt.
> de his divitiis sibi deducant dracumam, reddant cetera.'

> Now I will affirm that I do not recognise lot-drawers, nor those who prophesy for the sake of money, not even necromancers whom your friend Appius was accustomed to consult; in short, I do not care a fig for the Marsian augur, nor the neighbourhood soothsayers, **nor the astrologers around the Circus**, nor the guess-workers of Isis nor the interpreters of dreams. For these people are not diviners either by science or technique; instead
>
> 'they are superstitious seers and shameless prophets,
> either lacking in skills, mad or controlled by need,
> men who do not know the path themselves, and yet show the way to another;
> men who promise riches to others, and yet seek from them a drachma for themselves.
> From the riches that they predict, let them deduct a drachma and hand over the rest.'

Contempt towards these practitioners—including the astrologers—stems from the fact that they are unskilled amateurs who peddle their trade for financial gain and serve the individual interests of the rabble rather than the elite. This view is expressed elsewhere by elite Roman

[20] For the air of intellectual seriousness given to proper astrology, cf. 1.2 *scientiam*; 36 *numeris*; 91 *sollertiaque ingeniorum*. The strongly deterministic form of astrology is implied by 1.2 *scientiam . . . ut praedici posset quid cuique eventurum et quo quisque fato natus esset.*

writers and is in no way inconsistent with Quintus' earlier championing of proper, scientific forms of divination from both Rome and abroad.[21]

In summary, it is noteworthy that Cicero has fashioned his discussion in such a way as to emphasise the foreignness of proper astrology; even the potential enthusiasm of Quintus is toned down to such a degree that he is not allowed to make any connection between astrology and prominent Romans to bolster his case for the efficacy of divination. This might, I suggest, be taken as a reflection of the conservative elite position on astrology in Roman politics at the time of writing: while it might have attracted some elite enthusiasts, astrology had not yet achieved the sort of official acclamation which would have merited its inclusion among more long-standing forms of Roman divination.

The portrayal of astrology as foreign has the distinct advantage of allowing Marcus to be particularly direct and penetrating in his subsequent attack on the practice. There is no danger of his comments being perceived as a direct affront to Roman religious practice, so there is no need for him to resort to the sort of careful delineation between public, social duty and private, philosophical viewpoint that he feels he needs to make before attacking forms of Roman divination. Take, for example, the art of haruspicy (2.28):

> ut ordiar ab haruspicina, quam ego rei p. causa communisque religionis colendam censeo—sed soli sumus: licet verum exquirere sine invidia, mihi praesertim de plerisque dubitanti.

> I shall begin with soothsaying, which I judge should be cultivated for the sake of the Republic and a state religion—but we are alone: we have licence to investigate the truth without ill-will, especially in my case, as I am in doubt in many areas.

The directness of Marcus' attack against astrology can be felt in the contemptuous tone adopted towards its practitioners and the impassioned nature of his exclamations.[22] Moreover, in marked contrast to

[21] For peddlers of divination to the masses in Rome, cf. also Hor. *Sat.* 1.6.113–14 (*fallacem Circum*), Liv. 39.16.8, and the vivid picture presented in Juv. 6.582–91 esp. 585–6; see also Le Boeuffle (1989) 111–12, Bakhouche (2002) 138–9. For the elite attitude towards these low-class religious individuals, see further Nice (2001) 159–66, though I would disagree with Nice that foreignness per se plays a part in Quintus' negative assessment of these diviners. For a recent reassessment of the Roman 'street astrologer', see Ripat (2011), esp. 123–8.

[22] For the derogatory tone, see especially the use of *iste* at 2.88, 91, 92. Of the four exclamations from Marcus which begin with *o* (2.30, 90, 99, 103), two come from the section on astrology (2.90, 99).

the somewhat conciliatory tone adopted in the discussion of Roman divination, practices which are credited to human error or folly, Marcus attacks astrology as a deliberate act of deceit and the product of foolishness or madness.[23] It would appear that Marcus has picked up on Quintus' contempt for the vulgar astrologers (1.132), only to apply it to *all* astrological investigation.

Nor is the argumentation any less devastating than its tone. Marcus first constructs the position of the Chaldeans, namely 'hard' astrology, whereby the stars influence every aspect of a human's form and life (2.89). He then takes pleasure in knocking it down with no fewer than twelve forms of argument.[24] Marcus starts with the popular objection stemming from the case of twins who share the same birth time and physical features and yet fare differently in life (2.90).[25] He turns next to the fallibility of human sight in discerning the movements of planets so distanced from the earth (2.91). Next, stemming from an apparent belief that astrology takes no account of differing geographical position from which to view the stars, Marcus ridicules the assumption that the same star rises at a given time over all lands simultaneously (2.92–3). Moving at a faster pace, a swift succession of arguments criticises astrologers for taking no account of the influence of different types of weather (2.94); genetics and parenting (2.94); the fact that people born at the same moment have different lives, an extension of the earlier 'twins' argument (2.95); subsequent improvements that can be made to natal defects by means of medicine or training (2.96); (again) geographical variation (2.96–7). Next, and as the logical inverse of the 'twins' argumentation, Marcus notes that people born under different configurations of stars can share the same death, adducing the Roman example of Cannae as proof (2.97). He then asks

[23] For the deceit of astrology cf. esp. *fallacissimo* (2.91). For consistent accusations of foolishness/madness, cf. *o delirationem incredibilem* (2.90); *dementia* (2.94); *estne sanorum hominum* (2.94); *quid potest dici absurdius* (2.98). For the use of the more conciliatory term *error* to refer to other forms of divination, cf. 2.55, 83, 119, 120. Astrology sometimes also attracts this label, but typically in phrases which express the grave extent of the error; cf. *non enim omnis error stultitia dicenda est* (2.90); *mediocris erroris est?* (2.94); *o vim maxumam erroris* (2.99).

[24] For detailed critique of the philosophical argumentation of Marcus' attack on astrology, which draws on Greek sources (probably Panaetius) as well as the author's own inspiration, see esp. Long (1982) 172–8.

[25] For this popular attack, cf. Pers. 6.18–19 and Sen. *Ben.* 7.1.5; also August. *CD* 5.2–6. Cicero's friend Nigidius Figulus apparently countered such arguments by pointing to the tiny variations in heaven between the times of birth of each twin (August. *CD* 5.3). If he knew Nigidius' counterargument, Cicero makes no such concession here.

why only human beings, and not also beasts, are affected by astrology (2.98), and this leads him to the only acknowledgment in the entire work of a Roman practising astrology.

Lucius Tarutius Firmanus, a contemporary and friend of Cicero,[26] apparently used astrological calculations to ascertain the constellation under which the city of Rome was 'born'. Though commentators both ancient and modern have tended to be charitable towards Tarutius,[27] Cicero clearly sets this man up for light ridicule—2.99 *o vim maxumam erroris*—for trying to extend the (already suspect) power of natal stars to the foundation of a city. Furthermore, he can do so without compromising the strong distinction between astrological activity and proper Roman practice. Tarutius' horoscope of Rome held no official status or public significance: it was, as Plutarch intimates (*Rom.* 12), the result of a private intellectual request made by Varro to his friend, probably for inclusion in one of his (now lost) works. Moreover, Tarutius might be plausibly depicted as 'foreign' to Rome. First, as his name would suggest, he was most likely a native of the Latin town of Firmum in Picenum, and the cognomen might indicate a continuing strong connection with his native town.[28] Moreover, he apparently wrote in Greek, which further distances him conceptually from the city.[29] Cicero takes advantage of this to emphasise the 'foreignness' of Tarutius' activities: most notably, Cicero's reference to Libra as 'the yoke' (2.98 *in iugo cum esset luna*) consciously reproduces the Greek ζυγός, a term for the constellation unparalleled elsewhere in Latin.[30]

Reverting to the attack on astrology, Marcus reserves possibly the simplest and most effective argument until last: major, high-profile Roman horoscopes, involving Pompey, Crassus and Caesar, are refuted on a daily basis (2.99)!

In summary, I would argue that Cicero's treatment of astrology in *De Divinatione* can make a useful contribution to our understanding

[26] Cf. *Div.* 2.98 *familiaris noster*. He was also apparently a good friend of Varro (Plu. *Rom.* 12).

[27] Solinus, *c.* AD 200, refers to him as *mathematicorum nobilissimus* (1.18). For modern enthusiasm for Tarutius' astrological competence, see Grafton and Swerdlow (1986).

[28] See Rawson (1985) 34.

[29] See the catalogue of influential authors on Plin. *Nat.* 18 (Le Bonniec and Le Boeuffle 1972: 27–8). See further Rawson (1985) 64–5.

[30] See Bouché-Leclercq (1899) 141–2.

of the socio-political status of astrology in Rome before the appearance of the Julian comet. In the spirit of the new age of theological discourse, Cicero subjects a 'hard' conception of astrology to the rigours of Greek philosophical argumentation, and he can do so all the more vehemently because he recognises it as an entirely foreign and unofficial practice in comparison with the many officially sanctioned forms of Roman divination covered in the work.

ASTROLOGY IN *DE FATO*: A STELLAR COMPROMISE?

But let us turn now to *De Fato*. Cicero's now fragmentary treatise on fate was conceived as the third part of the author's examination of the philosophy of Roman religion (cf. *Div.* 2.3), and it was written (and one assumes published) not long after *De Divinatione*. At the beginning of the work, the author informs us that he had intended this treatise, like many of his previous philosophical enquiries, to take the form of a dialogue in which different sides of the argument are presented (*Fat.* 1). The dramatic setting for the treatise would certainly have been conducive to this: a meeting, at Cicero's villa in Puteoli, between himself and the consul-designate, Aulus Hirtius, whom Cicero portrays as being equally immersed in philosophy (*Fat.* 2). In reality, however, the treatise turns out to be an exposition by the author, rather than a dialogue between 'himself' and his guest, a simplification in style which Cicero suggests was forced upon him by 'a certain circumstance/misfortune' (*casus quidam*, *Fat.* 1). Fortunately, we are in a good position to fill in the political backdrop to the 'circumstance' affecting the author.

General date parameters can be confidently established for *De Fato*. The introduction places the dramatic date of the text at some point after the death of Julius Caesar (*post interitum Caesaris*, *Fat.* 2), and we know that the author died at the end of the following year. On this evidence alone, we can establish parameters of March 44 BC to December 43 BC. Furthermore, reference to Hirtius as *consul designatus* (*Fat.* 2) places the dramatic date of the meeting between Cicero and Hirtius before 1 January 43 BC, and it may be fair to assume that the composition date is also prior to that date, after which time such a label for Hirtius would have been redundant. This still leaves us with a nine-month period (March 44 BC—January 43 BC) in which to place the writing of the treatise, amid the hectic events in Cicero's personal life in a year marked by turbulence in the late Republic.[31]

The fairly precise dating posited by Yon (1944: ii–v) has so far found little opposition.[32] Cicero informs us (*Att.* 14.12.2) that he spent time at his villa in Puteoli (from 17 April to 23 May 44 BC) during which time Hirtius came to listen to him declaim. This chimes perfectly with the dramatic setting for *De Fato*. As for the time of writing, Yon pieces together what we know of Cicero's political and literary activities during this busy year to argue that the treatise must in all likelihood have been composed during May–June 44 BC, probably before July, in which Cicero was busy composing another treatise (*On Glory*) and setting out on a journey to Greece; after this time, Yon argues, Cicero became heavily engaged in writing political speeches against Antony (*Philippics* 1 and 2, Sep–Oct 44 BC) and in politics more generally at Rome.[33]

While I accept the dramatic date of the work, I see nothing in the text to prevent it from having been written over a longer period than that posited by Yon. First, considering the speed with which Cicero was able to compose different works of literature simultaneously in these latter years, it would be unwise to base an argument too heavily on a perceived lack of time on the author's part.[34] Secondly, and more positively, there is one particularly interesting section of *De Fato*—the brief discussion of astrology—which would, I argue, make a lot more sense if it were written in the aftermath of Caesar's comet in July 44 BC.

At *Div.* 2.89, Marcus had set up a strictly 'hard' definition of fatalistic astrology with the intention of knocking it down in various contemptuous ways. The approach to astrology adopted by Cicero in *De Fato* is quite different (*Fat.* 7–8):

[31] For general discussions of this period of Cicero's life, see, along with the discussion later, Rawson (1975) 260–77; Lacey (1978) 141–59; Lintott (2008) 339–407, 440–5.

[32] It is followed by Sharples (1991: 5–6) in his edition of the text.

[33] For *Philippics* 2 completed by the end of October, cf. Cic. *Att.* 15.13; for *De Gloria* mentioned or alluded to in letters dating from 11–25 July, cf. Cic. *Att.* 16.2.6, 16.3.1, 16.6.4.

[34] For the extensive list of Cicero's outputs in 46–44 BC, see e.g. May (2002) 20–1 (a list which does not even include the huge number of letters generated during this period). The ease with which Cicero could produce literature is further suggested by his apparent ability to compose *Topica* while at sea (*Top.* 5).

> inter locorum naturas quantum intersit, videmus; alios esse salubris alios pestilentis, in aliis esse pituitosos et quasi redundantis, in aliis exsiccatos atque aridos; multaque sunt alia quae inter locum et locum plurimum differant. Athenis tenue caelum, ex quo etiam acutiores putantur Attici, crassum Thebis, itaque pingues Thebani et valentes. tamen neque illud tenue caelum efficiet ut aut Zenonem quis aut Arcesilam aut Theophrastum audiat, neque crassum ut Nemea potius quam Isthmo victoriam petat. diiunge longius. quid enim loci natura adferre potest, ut in porticu Pompei potius quam in campo ambulemus, tecum quam cum alio, Idibus potius quam Kalendis? ut igitur ad quasdam res natura loci pertinet aliquid, ad quasdam autem nihil, **sic astrorum adfectio valeat, si vis, ad quasdam res**, ad omnis certe non valebit.
>
> We can see the extent of the difference between the nature in different places: that some are healthy, others plague-ridden, in some places men are phlegmatic as if overflowing with moisture, in others they are dried up and parched—and there are many other things besides which differ significantly between one place and another. At Athens the air is thin, because of which the people of Attica are even thought to be more sharp-witted, while at Thebes it is dense, and so the Thebans are dull and strong. Nevertheless, that thin air will not compel anyone to listen to Zeno, Arcesilaus or Theophrastus, nor will dense air compel anyone to seek victory at Nemea rather than at the Isthmus. To continue this distinction further: what power can the nature of a place bring to bear such that I walk in Pompey's portico rather than in the Campus, with you rather than with someone else, on the Ides rather than on the Kalends? Therefore, just as the nature of a place has some bearing on some things, but nothing on certain other things, **so too let the influence of the stars have power over some things, if you like**: it certainly will not have power over all things.

Taking a cue from climatic conditions, Cicero appears to concede that the stars may exert some influence on mortals, but not that they determine every aspect of their appearance or action. He makes the case in the most pointed way possible to his Roman addressee by alluding to the recent death of Caesar: by inference, the stars may have exerted some influence on the dictator's life, but they did not determine his walking in Pompey's portico with Hirtius on the Ides of March. In effect, Cicero is the first Roman to articulate a weaker brand of influence for the stars on human life.[35] This is all the more surprising in the light of his rejection of a similarly soft approach, advocated by Diogenes the Stoic, at *Div.* 2.90–1:

[35] For the various layers of non-determinist approaches to astrology in the Roman world, see Lehoux (2012) 159–75.

quibus etiam Diogenes Stoicus concedit aliquid, ut praedicere possint dumtaxat qualis quisque natura et ad quam quisque maxume rem aptus futurus sit; cetera quae profiteantur negat ullo modo posse sciri . . . at ego id ipsum quod vir optumus Diogenes Chaldaeis quasi quadam praevaricatione concedit, nego posse intellegi.

[To the Chaldeans who hold a notion of hard astrology] even the Stoic Diogenes makes some concession: he says that they have the power to make predictions insofar as they can determine the disposition of each person, and to which activities each person will be best suited; as for all the other things that they predict Diogenes denies that they can know this by any means . . . But I for my part say that even that concession which our excellent friend Diogenes makes to the Chaldeans as if by some form of collusion cannot be comprehended.

If Cicero's approach to astrology in *De Fato* does not amount to a complete change of philosophical position from *De Divinatione*—after all, strongly determinist astrology is still ruled out[36]—it is at least a subsequent *softening* or *clarification of* the argument as presented in his previous treatise, which ruled out any sort of 'collusion' (*Div.* 2.91 *praevaricatione*) with astrological influence. To accompany this change in stance on astrology, there would also appear to be an elevation in status for its practitioners in *De Fato*, as the deceitful foreign conmen of *De Divinatione* are replaced by individuals who, indistinguishable from other diviners (*Fat.* 15 *Chaldaeos ceterosque divinos*), are at home within a Roman philosophical debate (*Fat.* 11–16).

This change in stance on astrology over the space of two works and over the timeframe of a few months has rarely been felt worthy of note, let alone explanation.[37] As with the apparent shifting of philosophical positions elsewhere between Ciceronian works, the present change might be explained in terms of the Academic philosopher's freedom to pick up and put down theoretical positions at will, and as a further manifestation of the new age of religious enquiry. I would, however, advance a more politically oriented proposition: that this may be a strategic reaction to the fast-growing importance of astrology in the wake of Caesar's comet.

[36] Lintott (2008: 358) reads *De Fato* as a re-enforcement of the argument against the brand of strong determinism found in *De Divinatione*, at a time when Cicero wants to promote the view that the demise of the Republic is not inevitable.

[37] The difference is noted in passing by Guillaumont (1984: 162 n. 12), but no explanation is offered. Schofield (1986: 50) draws attention to the acute change in methodology and style between *Div.* and *Fat.*—from examination of concrete examples to intricate and abstract philosophical argumentation—but has nothing to say about the shift in emphasis on fatalistic astrology.

Scholars of *De Fato* have, not surprisingly, tended to concentrate on its philosophical density.[38] But this approach runs the risk of overlooking the highly charged political atmosphere of crisis which forms the backdrop (*Fat.* 2–3):

> nam cum essem in Puteolano Hirtiusque noster consul designatus isdem in locis, vir nobis amicissimus et his studiis in quibus nos a pueritia viximus deditus, multum una eramus, maxime nos quidem exquirentes ea consilia quae ad pacem et ad concordiam civium pertinerent. cum enim omnes post interitum Caesaris novarum perturbationum causae quaeri viderentur iisque esse occurrendum putaremus, omnis fere nostra in his deliberationibus consumebatur oratio, idque et saepe alias et quodam liberiore quam solebat et magis vacuo ab interventoribus die. cum ad me ille venisset, primo ea quae erant cotidiana et quasi legitima nobis, de pace et de otio; quibus actis 'quid ergo?' inquit ille 'quoniam oratorias exercitationes non tu quidem ut spero reliquisti, sed certe philosophiam illis anteposuisti, possumne aliquid audire?'

> For when I was at my villa in Puteoli and my Hirtius, consul designate, was in the same region—a man who is most friendly towards me and is devoted to these (philosophical) pursuits in which I have spent my life since boyhood—we were together a great deal, especially for our part thrashing out the sort of plans which might bring peace and harmony among the citizens. Following the death of Caesar, it seemed that every reason was being sought to create new disturbances, and we thought that we should counter them: for this reason, practically all our conversation was taken up in deliberations of this kind. Amidst numerous other meetings, there was one particular day which allowed more leisure time than usual and was freer from interruptions. When he had come to me, at first we spoke about those things which were our daily concern and appropriate for us, as it were, namely peace and tranquillity. When we had done this, he said, 'Well, since you indeed have not, I hope, abandoned rhetorical exercises, though you have certainly given philosophy priority over all of them, can I hear something from you?'

De Divinatione was presented as a private and frank conversation between brothers, at a time when Cicero's retirement from politics and the public sphere afforded an opportunity for such discussion—all of which helps to explain the 'unguarded' nature of Marcus' attack on forms of state divination. *De Fato* is advertised as something quite different, since it is set firmly in a contemporary political context and

[38] See Schofield (1986: 50): 'it is the Ciceronian treatise philosophers most enjoy reading'. For detailed discussion of the philosophy of *De Fato* 7–9, see Sedley (1993).

features a pro-Caesarian interlocutor. After the death of Caesar, Cicero again became actively involved in matters of state, with only fleeting opportunities for his former philosophical pursuits. More significantly, he is conversing with—and the philosophical discussion is sanctioned by—Aulus Hirtius, the former officer of Caesar whom the dictator had entrusted to add the eighth book to his *Gallic War*. Hirtius' continuing loyalty to the memory of Caesar, a cause quickly taken up by Octavian, is shown not only by the gushing praise he heaps on Caesar's career and writing ability,[39] but also by his participation in Octavian's war at Mutina against Antony in 43 BC, during which Hirtius lost his life.

It is this contemporary political context, at a time when an astrological interpretation of Caesar's comet has gained currency, which, I think, offers the best explanation for Cicero's changing or softening of philosophical stance towards astrology and its practitioners in *De Fato*.

Scholars have often been surprised that Cicero makes no reference at all to Caesar's comet in his surviving writings. Various possible reasons have been offered to account for this omission. From 17 July to 1 August, Cicero was off the coast of southern Italy during the time of the comet, which may not have been visible from that location. He might have felt it too blatant a contradiction to champion a predictive quality to the Caesarian comet when he had only recently dismissed such a notion (*Div.* 2.60). One might also point to the paucity of the letters that have survived from this period, as well as the selectivity of the material in the few extant letters.[40] While these reasons may explain why Cicero avoided discussion of the event itself, scholars are right to remain troubled by Cicero's silence on the political impact of the comet in his private correspondence, especially given his newly invigorated political career.[41] *De Fato*, I suggest, bridges the gap: a strategic repositioning/softening of stance enables Cicero to bring himself in line with the fast-moving political importance of astrology without having to endorse directly the popular interpretation of the comet, which would have struck him as philosophically unsophisticated.[42]

[39] Cf. Caes. *BG* 8 *praef.* 5–6; Suet. *Iul.* 56

[40] For discussion of the various possible reasons for Cicero's silence on the comet of Caesar, see Ramsey and Licht (1997) 95–6, 112–16.

[41] For his general recording of interest in other games and public entertainment in that year, cf. e.g. *Att.* 14.2; *Phil.* 1.36; *Att.* 15.26, 15.28 (Apollinarian Games, 6–13 July).

[42] To be sure, the curt phrase *post interitum Caesaris* (*Fat.* 2) alerts the reader early on to the fact that Cicero will not be paying any lip service to popular rumour about Caesar's stellar trajectory.

Such a conciliatory strategy would be entirely in line with what we know of Cicero's public stance towards Caesar in the six months following his assassination. Cicero quickly fashioned himself as an advocate of compromise between pro- and anti-Caesarian factions, inviting his fellow Romans to rise above personal differences in the interests of peace and harmony of the State. He first set out his political stall in a meeting of the Senate a few days after Caesar's death, on 17 March, in which proposals for amnesty to the conspirators were accompanied by calls for a decree recognising the validity of Caesar's laws and decrees.[43] Cicero's strategic positioning finds greatest expression in *Philippics* 1, a speech delivered to the Senate on 2 September, in which the orator reinforces the line that Caesar's *acta* should be upheld—in the face of apparent abuse of Caesar's legacy by Antony—while at the same time condemning Caesar's version of dictatorship and arguing against Antony's proposal to add to all thanksgivings a day in honour of Caesar.[44]

If Antony's religious proposal had been motivated in any way by the recent popular belief in Caesar's astral destiny,[45] Cicero was keen to pull back from moves which compromised the traditional Roman religious segregation between thanksgiving and the honouring of the dead (the Parentalia) without causing undue dishonour to the late dictator.

The new approach towards astrology adopted in *De Fato* would harmonise well with Cicero's professed commitment to public reconciliation and his public stance towards the memory of Caesar post July 44 BC. Cicero's endorsement of the astrological underpinning to the comet—that the stars exert some influence on mortals—does just enough to accommodate popular interpretations of Caesar's destiny. It does not, however, challenge traditional Roman religious practice, nor does it raise an overtly pro-Caesarian chorus by accepting wholesale the popular interpretation of the comet.

In fact, the 'soft' approach to astrology adopted in *De Fato* might be felt to provide the ideal political compromise: it allows one to flatter

[43] Cf. Dio 44.22.2–34.1, esp. 44.34; Vell. 2.58.4 (amnesty of assassins); App. 2.135 (upholding of *acta*).

[44] For the general context and structure of *Philippics* 1, see Ramsey (2003) 81–4. Antony's proposal to add an extra day to every thanksgiving is implied by *Phil.* 1.12–13 and 2.110; for Cicero's response, cf. *Phil.* 1.13. For Cicero's defending Caesar's *acta*, cf. *Phil.* 1.16–26. For Cicero's condemnation of Caesar's dictatorship, cf. e.g. *Phil.* 1.4, 9.

[45] As, for example, thinks Ramsey (2003: 113).

the individual with a notion of divine stellar favour without taking away individual (earthly) achievement. In adopting this line, it is possible that Cicero was now aligning himself with the views of the late dictator himself. Caesar's general views on astrology may be gleaned, with caution, from his surviving works. Caesar's advocacy of the power of the stars over human life may be constructed from his work, *De Astris*, but it was a 'soft' approach at best: most notably, the near absence in his war commentaries of detail about any communications he may have made with the gods, especially in the form of augury and sacrifice, illustrates the premium he placed on his own free will and his own mortal achievements.[46] This is exactly the refined position of Cicero in *De Fato*: some astrological influence is conceded, but any divinatory theory which impinges on free will is promptly discarded.

To summarise, Beard may well be right to suggest that Cicero was the pioneer of religious discourse at Rome. But we must remember that he achieved this at a time when he was free from politics and possibly not expecting to re-enter the arena. As soon as he was unexpectedly thrust once more into the political limelight, theological discourse had to be tied to fast-moving political posturing towards divination. Cicero may mark the beginning of religious discussion on astrology, but he may also be responsible for cutting it short. The removal of the second speaker from *De Fato*, then, is symptomatic of a sudden curtailing of religious debate on astrology: no one steps forward to challenge the views expressed in *De Fato* 7–8. One would have to wait 150 years, for the mercurial Favorinus, for a sustained philosophical attack on astrology to match the vehemence of Cicero's in *De Divinatione*.[47] In the meantime, Cicero had laid down for subsequent authors the first steps of a discourse on astrology, one that keeps a close watch on the developing political strategies of Octavian/Augustus. This would become all the more important for authors operating in the Augustan age itself.

[46] For the *De Astris*, see Chapter 8. For Caesar's general downplaying of divine involvement in his own achievements in his *commentarii*, see e.g. Wardle (2009) 107–8.

[47] See Chapter 11.

Part 3

Astrology for the Augustan Age

5

To Have and to Hold

Astrology for an Imperial Age

In order to provide more detailed context for the Augustan authors discussed in later chapters, it is important to set out, briefly, Augustus' special brand of astrological celebration and discretion during his reign.

CAPRICORN: A SIGN FOR A DYNAMIC EMPEROR

As we have mentioned (Chapter 3), Octavian's interest in genethlialogical astrology may be traced back at least as far as 44 BC. Moreover, in my opinion, Octavian/Augustus published his horoscope early, and certainly by the late 20's BC. This enthusiasm for his own personal astrology soon manifested itself in the visual space of Rome. Suetonius informs us (*Aug.* 5.1) that Octavian was born nine days (inclusive counting) before the Kalends of October. However, even though it was neither the sun-sign nor the Ascendant for his horoscope, Suetonius (*Aug.* 94.12) and the extant record confirm that it was Capricorn which was taken up by Octavian as the dominant astrological symbol of his career.[1]

[1] There are examples which may be advanced to argue for a recognised connection between Octavian/Augustus and Libra: Virgil (*G.* 1.32–5) suggests that a space will be made available for Augustus between the constellations of Scorpio and Virgo (i.e. the place occupied by Libra), and both Manilius (4.542–6, 773–7) and *De Duodecim Signis* (*Asclepiadii*) (*Anthologia Latina* I.2.618.4 Buecheler/Riese) refer to Libra in connection with an imperial ruler. But one must be cautious: Virgil could just as easily be referring to Julius Caesar; the passages from Manilius' fourth book may well be addressed to Tiberius; and there is no telling to which ruler the generic 'Caesar' of the anonymous poem refers; see further Barton (1995: 35–6) and Lewis (2008: 313–21). In spite of these apparent complications, therefore, Capricorn may well have emerged as an unchallenged astrological symbol for Octavian/Augustus (though see also Lewis (2008: 330–4) for a proposed connection between the Emperor and Virgo through Augustan iconography).

The reason for Octavian's adoption of this symbol has intrigued scholars, and various theories have gained currency.[2] Explanations from the strictly astrological sphere were the first to be entertained: rather than a sun-sign at his birth, perhaps Capricorn represents the sign at Octavian's *conception*, or perhaps his *moon*-sign was in Capricorn. More imaginative astrological theories have also been touted. Perhaps it is significant that Capricorn was in the ascendant at the time of Caesar's comet (20–23 July): did this encourage Octavian to adopt it as his sign, treating the comet as some form of second birth?[3] Or, working from the astrological system which suggests that the Lot of Fortune falls in Capricorn, was this constellation felt to carry a positive auspicious significance?[4]

This last theory moves us from the strictly astrological to a more distinctly ideological quality behind the symbol, and it is the rich ideological potential of Capricorn which undoubtedly secures its endurance for Augustus at different phases of his career. The first secure dating for an official representation of Capricorn is on a denarius from the eastern mint of 28 BC, which celebrates the victory against Egypt at Actium.[5] Capricorn is an especially apt symbol in this regard. First, it is a sign which, like Augustus, is felt to hold particular sway over the west.[6] Secondly, Capricorn's amphibious nature might have provided a convenient means of symbolising Octavian's dexterity over both land and sea in post-Actium rhetoric. Furthermore, the figure of Capricorn is identified with Pan, in his role as restorer of the rule of the gods against Typhon, and the Egyptian Horus, who committed vengeance on Seth for the murder of his father, Osiris.[7] Fittingly, then, Augustus' astrological symbol is charged with ideas of restoration of divine order and righteous vengeance, and its strong

[2] For the history of the debate and detailed discussion of many of the options, see esp. Barton (1995). See also Kraft (1967); Dwyer (1973); Barton (1994a) 40–7; Ramsey and Licht (1997) 147–53; Gee (2000) 140–2; Rehak (2006) 72–4.

[3] See Ramsey and Licht (1997) 147–53.

[4] For the Lots, cf. Man. 3.160–202 with Bouché-Leclercq (1899) 296–9 and Barton (1994b) 98–9.

[5] For a concise discussion of the visual representations of Capricorn during the career of Octavian/Augustus, see Barton (1995) 48–51.

[6] See esp. Man. 4.791–6; alluded to also at Hor. *Carm.* 2.17.19–20 (*tyrannus Hesperiae Capricornus undae*) and Prop. 4.1B.86 (*lotus et Hesperia quid Capricornus aqua*).

[7] The connection between Capricorn and both Pan and Egypt is made by Nigidius Figulus; see esp. Dwyer (1973) 61–6.

military connotations made it an ideal emblem for the standards of many Roman legions.[8]

The symbol crops up again around the time of the (diplomatic) recovery of the Roman standards from the Parthians—the coinage dates from 20 to18 BC—where notions of restoration of order and righteous vengeance would again have been pertinent. Looking forwards to the more positive rhetoric of the Augustan age as an era of restored peace and prosperity, Capricorn proves to be equally versatile. As both the new sun (winter solstice) and the New Year take place during Capricorn, the sign is particularly appropriate for expressing the dawn of a new era.[9] Furthermore, in association with the cornucopia, the Horn of Plenty, Capricorn is able to contribute to the rhetoric of a restored Golden Age of fertility and abundance.[10]

Scholars are right to conclude, therefore, that Capricorn's enduring symbolic power stems from its flexibility and multiplicity of association, its ability to improvise through time to match the requirement of the moment.[11] The power of this imperial symbol is such that it finds itself represented in private contexts, from cheap terracotta antefixes to expensive gems, which may suggest dissemination through different classes eager to show their loyalty to the Emperor.[12] Moreover, whatever its original astrological significance, it is the ideological charge of Capricorn that

[8] For Capricorn on Augustan legionary standards, see Keppie (1984) 139–40. The specific motif in the Horus myth of rightful vengeance against a murdered *father* does, of course, make a particularly strong link to Octavian's actions and rhetoric post 44 BC, and it is possible that visual associations between Octavian and Capricorn go back to this period, if a coin minted by the prefect Q. Oppius can be dated to 41/40 BC; for the dispute over dating this coin, however, see Kraft (1967); Barton (1995) 48.

[9] Capricorn's association with the new sun would also provide fitting symbolic affinity with another of Octavian's much-vaunted images, the sun-god Apollo; see Zanker (1988) 48–53. The importance of Capricorn as herald of the New Year might also have been represented visually in the length of the shadow cast by the Horologium Augusti (on which see later).

[10] See esp. Galinsky (1996) 111–17; Gee (2000) 133–42.

[11] See Barton (1995) 47.

[12] See esp. Barton (1994a) 42–4. The most famous example is the Gemma Augustea, which features Capricorn inscribed on a solar disc next to Augustus' head; Jeppesen (1994: 347) argues plausibly for a dating of AD 9–10, and suggests that the gem was commissioned by Augustus himself; for debate, see also Zanker (1988) 230–1; Pollini (1993) 280–5. Perhaps surprisingly, the figure of Capricorn did not prove attractive to contemporary authors as a mythical space in which to promote imperial astrological symbolism. Germanicus gives Capricorn only brief, albeit respectful, treatment in his *Phaenomena* (554–60; see Chapter 8), and Ovid deals briefly and subtly with the imperial symbolism behind Capricorn via his treatment of Capella, the 'Little Goat'; see Gee (2000) 127–42, 152–3.

endures, as future Emperors, though born at different times of the year, maintain Capricorn as a key symbol of their imperial identity.[13]

THE PANTHEON AND THE SUNDIAL: ASTROLOGY IN ARCHITECTURE?

The presence of coins with positive imperial astrological symbolism, dating from at least 28 BC, demonstrates the perceived visual power of astrology to articulate political messages. But did Augustus also conceive of grander visual representations of his own astrology? The question has often been asked with regard to two specific structures of his reign—the Pantheon and the Horologium—to which we now turn.

The Pantheon was built by Agrippa from 27–25 BC.[14] Our fullest ancient testimony comes from Cassius Dio (53.27.2–3), who speaks of a structure which contained statues of both gods and mortals.[15] It would appear, therefore, that it was not so much a temple for all the gods, as the name might suggest, as a monument to the *gens Iulia* celebrating the imperial family and its divine ancestry.[16] But perhaps we can read something more celestial into the structure's symbolism, as prompted by Dio himself (53.27.2):

> ... ὡς δὲ ἐγὼ νομίζω, ὅτι θολοειδὲς ὂν τῷ οὐρανῷ προσέοικεν.
>
> ... as I believe, it [i.e. the name] is because it resembles the heavens by virtue of its vaulted roof.

Is Dio right to suggest that the dome's interior was intended to represent the heavenly sphere?[17] If so, then perhaps other celestial themes

[13] For references, see Barton (1995) 47.

[14] For modern discussions of the Pantheon, see Ziolkowski in Steinby (1993–2000) 4.54–61; Wilson Jones (2000) 177–213.

[15] Venus, Mars, Julius Caesar and Augustus are mentioned specifically. It is unclear whether the statue of Caesar showed him as a god or, perhaps more likely, in mortal form; see Fishwick (1992). It would have been most out of touch with Augustan visual culture, however, to have presented Augustus in anything other than mortal form.

[16] Mars might seem a little out of place, but he might be justly included as father of Augustus' ancestor Romulus.

[17] For open-minded assessments of the celestial overtones, see MacDonald (1976) 77, 89; Loercke (1990); Wilson Jones (2000) 182–4; Thomas (2004) 28; Lancaster (2005) 46, 174–5; more dismissively, see Ziolkowski in Steinby (1993–2000) 4.60–1. Joost-Gaugier (1998) also discusses cosmological associations in the Pantheon, but takes it to be an essentially Hadrianic design.

can be traced in the Pantheon's design. It is noticeable that the only two gods to be mentioned—Mars and Venus—are also planets, and one might put this together with the existence of the seven exedrae within the monument to argue that this structure may have specifically housed the guardian deities of the seven planets. Moreover, the slightly convex nature of the rotunda floor might be felt to represent the surface of the earth. Unfortunately, Dio's tantalising extract is marred by the fact that the Pantheon witnessed by the author was *not* the original Agrippan structure, which had been destroyed by fire in AD 80 (Dio 66.24.2), but the one rebuilt/restored by Hadrian in AD 126.

As we have no way of knowing the extent to which the Hadrianic monument resembles the original Agrippan precedent, we cannot recover with any certainty the structure and the symbolism of the Augustan monument. For an Emperor who is prepared to advertise his astrology visually on coins, however, it would seem to be a logical step for him to have pursued this strategy with much larger structures. The Pantheon, built within three years of the first definite numismatic representation of Augustan astrology, might have been designed to symbolise harmony between the imperial family and the heavenly influences, as statues of the *gens Iulia* occupy the same space as planetary deities within a spherical, cosmic structure.

An obelisk, which stands today in the Piazza Monte citorio, originally formed the gnomon (shadow-caster) of a giant instrument referred to as the Horologium Augusti, which was brought from Egypt to Rome in 10 BC and placed on the Campus Martius in the proximity of both the Ara Pacis (built 13–9 BC) and the Mausoleum. The purpose of the structure has generated lively theories among scholars. Most famously, Buchner argued that the Horologium was a daily time-keeping device in the form of a giant sundial.[18] This was no neutral instrument, Buchner maintained, but a work of technological genius which promoted complex imperial discourse: not only was the obelisk oriented towards the Mausoleum, but the shadow cast by the obelisk pointed towards the Altar of Peace on Augustus' birthday. Buchner's theories have since been subject to some heavy criticism, and have given way to general consensus that the obelisk was in fact a solar meridian, an astrological instrument designed to chart the progress of the solar year through the signs of the zodiac.[19]

[18] His individual studies are collected in Buchner (1982).

[19] For heavy criticism of Buchner's theories, see esp. Schütz (1990); for modern supporters of Buchner, see e.g. Rehak (2006) 80–7. For the history of the debate, see Heslin (2007).

If current scholarly consensus denies the monument the sort of impressive astrologically charged ideology that Buchner had given it, it nevertheless remains a potent symbol for Augustus. Coming from the Egyptian city of Heliopolis, the city of the sun, the monument contributes to the Emperor's identification with the sun (through Apollo), as well as to a general post-Actium sense of Roman subjugation of Egypt. It also advertises Augustus' control over time, and might have fulfilled a symbolic function of showing that Augustus—who had recently taken charge of the Roman calendar on becoming *pontifex maximus* in 12 BC—was keen to establish a means of ensuring that the newly configured civic calendar was kept in synch with the solar year.[20] Moreover, if Pliny is correct when he says that the paving for the monument stopped at the point where the shadow cast by the gnomon reached the winter solstice (*Nat.* 36.72), a powerful visual connection between Capricorn and the beginning of a new era of time might have been made, at least by the learned viewer.[21] At the very least, the presence of the monument in the heart of the city attests to an Augustan desire to give the foreign practice of astrology visual recognition among more traditional religious buildings.

RESTRICTING UNSANCTIONED ASTROLOGY

Personal astrology, then, was emerging as a powerful tool in the political discourse of Octavian/Augustus. But the essential problem for Octavian was that a positive horoscope was not as securely exclusive to the Emperor as, for example, the rhetoric surrounding his divine ancestry. There were many astrologers in Rome and, therefore, as many horoscopes to be cast as there were Romans prepared to make use of their craft. And who was to say that another astrologer might not at another time provide an equally glowing, if not more glowing horoscope for an individual with ambitions of power?

AD 11 witnessed the most official demonstration of a double-standards policy from the Emperor towards astrology. In this year, Augustus both publicised his own horoscope and outlawed the use of natal astrology by others, a move which attracted cries of hypocrisy

[20] See esp. Heslin (2007) 5–6.
[21] See esp. Barton (1994a) 45–7; Barton (1995) 44–7.

from Cassius Dio.[22] This should be seen not as a new move, but as the visible tip of a political iceberg. The Emperor's strategies of promoting himself through astrology while at the same time restricting others from making use of its power can be seen very early in his career, perhaps as early as 33 BC, if we agree to an early date for the publicising of the horoscope:[23] at that time, the dissemination of Octavian's horoscope would have co-existed with an expulsion of astrologers from Rome.

Octavian realised the dangerous potential of astrology in 33 BC when he banned astrologers and magicians from the city of Rome. Not surprisingly, this initiative was motivated by anxieties surrounding leadership. The propaganda war with Antony in the East was reaching a peak, and both the oriental origin of many of the astrologers and their potential influence on the masses at this delicate time must have concerned Octavian. He took no chances and installed Agrippa as aedile, who simultaneously banned astrologers from Rome and courted the affection of the masses through handouts and shows.[24] This action may have been followed a few years later, in 28 BC, by the expulsion from Rome of a specific practitioner, Anaxilaus, the mercurial Thessalian among whose dark arts was probably astrology.[25]

The foundation of the principate in 27 BC seems to have ushered in a period of relative calm with regard to Augustus' anxieties towards astrologers. But the connection between political anxieties and unsolicited prophetic consultation would emerge again in 12 BC, the year in which Agrippa died and in which Augustus' problems in finding an imperial successor magnified. Taking advantage of his recent assumption of the post of *pontifex maximus*, Augustus conducted a detailed scrutiny of all prophetic writings in circulation (Suet. *Aug.* 31):[26]

> postquam vero pontificatum maximum, quem numquam vivo Lepido auferre sustinuerat, mortuo demum suscepit, quidquid fatidicorum librorum Graeci Latinique generis nullis vel parum idoneis auctoribus vulgo ferebatur, supra duo milia contracta undique cremavit ac solos retinuit Sibyllinos, hos quoque dilectu habito; condiditque duobus forulis auratis sub Palatini Apollinis basi.

[22] See Dio 56.25.5 with discussion later.

[23] As I have argued in Chapter 3.

[24] Dio 49.43.1–5; see further Cramer (1951) 17–21. For a radical reassessment of Roman expulsions of astrologers, however, see Ripat (2011).

[25] See Cramer (1954) 85–6.

[26] The text of Suetonius is taken from Ihm (1964).

> After Augustus had indeed at last taken up the position of *pontifex maximus* on the death of Lepidus—whom he had never seen fit to remove from the position while he was alive—he collected up whatever prophetic books there were anywhere in circulation in Greek and Latin, from authors unknown or of no sufficient authority, and burned more than two thousand of them; only the Sibylline books did he retain, and amongst these also he made a choice. He deposited them in two gilded cases under the pedestal of Palatine Apollo.

This is the first definite example of double standards on the part of the Emperor in matters of divination: imperial promotion of (presumably) positive prophetic material goes hand in hand with clamp down on unsolicited and unofficial predictions of the future. This complex is found specifically with relation to astrology in the Emperor's famous edict of AD 11: not simply applicable to the city, Augustus' empire-wide ruling criminalised all consultations about death—i.e. all genethliological astrology, since this mapped out one's life from birth to death—and all consultations held in private between diviner and client alone (Dio 56.25.5). This seems to have been accompanied, according to Dio, by official publication of his own horoscope, designed probably to counter rumours of the aged Emperor's imminent death.

It should be noted that the edicts against astrologers in 33 BC and AD 11 were, like all edicts, short-term measures, one year in duration, proclaimed at a time of particular tension.[27] Nevertheless, as I have suggested earlier, these edicts should not be used to suggest that imperial nervousness towards astrology was somehow sporadic. On the contrary, they represent official manifestations of a general feeling of unease towards unsolicited astrological consultation which occupied the mind of the Emperor from very early in his career.

THRASYLLUS AT THE COURTS OF AUGUSTUS AND TIBERIUS (AD 2–16): THE EMERGENCE OF THE IMPERIAL ASTROLOGER

The empire-wide edict of AD 11 occurred at a time when the first high-profile astrologer of the Roman world, the Greek Thrasyllus, had firmly established himself within the imperial court.[28] Thrasyllus'

[27] See Cramer (1951) 10–11.
[28] For the most detailed analysis of Thrasyllus, see Cramer (1954) 92–108.

entry into the world of imperial politics may have come about by pure chance, as he appears to have been at Rhodes at the time of Tiberius' self-imposed retirement to the island from 6 BC–AD 2. In the light of his apparent eminence in various fields of academia—numerology, philosophy and philology, as well as astrology—it may be prudent to suggest that Thrasyllus made the acquaintance of Tiberius in a scholarly capacity, perhaps as a teacher, rather than as a result of an astute astrological consultation.[29] The astrologer accompanied Tiberius back to Rome in AD 2, where he was probably rewarded with Roman citizenship—taking the name Tiberius Claudius Thrasyllus—and a foreign dignitary as a bride.

Thrasyllus' subsequent relationship with Augustus is more difficult to reconstruct. Suetonius (*Aug.* 98.4) provides us with anecdotal evidence for a jovial and convivial relationship between the two in the latter years of the Emperor's life, and it seems likely that this represents many years of friendship, stemming back to a period soon after Thrasyllus' arrival into Rome.[30] It is particularly tempting, therefore, to see the influence of Thrasyllus behind the Augustan edict of AD 11.

Nothing, however, could match the special and long-standing bond between Tiberius and Thrasyllus, and the accession of Tiberius in AD 14 naturally heralded an increase in stature for the astrologer. Probably within the first year of his reign, he secured for Thrasyllus further elite status: already a Roman citizen, the astrologer's daughter was now entered into matrimony with a Roman knight, L. Ennius. Then, in AD 16, Marcus Scribonius Libo Drusus became the first high-profile Roman to fall foul of the spirit of the Augustan edict of AD 11 when he was seduced into consulting a range of diviners, including astrologers, to enquire about his own political future at the expense of the Emperor. Though he committed suicide before he was arrested, the trial went ahead regardless and a guilty verdict was reached.[31] Tiberius acted swiftly to renew the official imperial stance towards unsolicited astrological consultation by expelling astrologers from both Rome

[29] The sources tell the story of a nervous Tiberius who consulted astrologers at the time through fear of his own death, and subsequently killed them through fear that they would reveal the contents of their consultations; only Thrasyllus survived by astutely predicting his own death, a prediction which, in the light of the Tiberian cull of astrologers, convinced the Roman of Thrasyllus' superior skill; for divergent versions of Tiberius' testing of Thrasyllus, cf. Tac. *Ann.* 6.20–1; Suet. *Tib.* 14.4; Dio 55.11.2–3. See also Krappe (1927); Cramer (1954) 92–4; Oliver (1980).

[30] See Cramer (1954) 98–9.

[31] The fullest version of this story is given by Tacitus (*Ann.* 2.27–32). See also Shotter (1972).

and Italy, possibly in the form of two decrees of the Senate.[32] Once again, one senses the influence of Thrasyllus behind Tiberius' public postures on astrology, and this imperial aide would continue his role until his death in AD 36.[33]

ASTROLOGY FOR AUGUSTUS: BELIEF OR POLITICAL EXPEDIENCY?

In summary, Augustus shows a consistent preoccupation with astrology during his life: the potentially widespread promotion of his personal astrology, in the form of coinage, architecture and publicised horoscope was simultaneously matched by a tangible sense of nervousness about the use of natal astrology by others, a nervousness which finds official expression in edicts published during times of particular challenge to his leadership.

In the light of this preoccupation, it would be very interesting to ascertain whether or not Augustus actually *believed* in astrology in either its strongly deterministic or 'softer' forms. As ever, such enquiries are fraught with peril, but his own testimony in *Res Gestae* might provide the glimmer of an answer.

Cooley (2009: 158–9) astutely notices a horoscopic tinge to the way the Emperor appears to mark his *birth*, as well as his assumption of the position of *princeps*, as a pivotal turning point in the fortunes of Rome (*RG* 13):[34]

> [Ianum] Quirin[um, quem cl]aussum ess[e maiores nostri voluer]unt, cum [p]er totum i[mperium po]puli Roma[ni terra marique es]set parta victoriis pax, cum, **pr[iusquam] nascerer**, [a condita] u[rb]e bis omnino clausum [f]uisse prodatur m[emori]ae, ter me princi[pe senat]us claudendum esse censui[t].

[32] See also Cramer (1951) 21–9.

[33] For this period—outside the chronological confines of the present study—see esp. Cramer (1954) 102–8. It was Tiberius' close association with Thrasyllus which gave rise to the posthumous anecdotes about the Emperor's 'flock of astrologers' (*cum grege Chaldaeo*, Juv. 10.94), nervously consulted to determine whether the horoscopes of political rivals presented a threat to him (Dio 57.19.3–4). Thrasyllus' hold on the imperial court might even have been felt after his death, if we accept the popular identification of Balbillus, astrologer to both Claudius and Nero, as his son; see Cramer (1954) 112–31.

[34] The text and translation of *Res Gestae* are taken from Cooley (2009).

> Our ancestors wanted Janus Quirinus to be closed when peace had been achieved by victories on land and sea throughout the whole empire of the Roman people; whereas, **before I was born**, it is recorded as having been closed twice in all from the foundation of the city, the senate decreed it should be closed three times when I was leader.

Any potential astrological charge apparent in this part of the text is, however, distinctly absent from *Res Gestae* as a whole, which makes no direct reference to any celestial apparatus—divine or stellar—which may have supported the Emperor in his achievements. One can certainly point to formal reasons for this silence in *Res Gestae*: influenced as it is by laudatory inscriptions and funeral orations, it is entirely appropriate that the focus is maintained on the earthly achievements of the individual.[35] But then Augustus is no ordinary individual, and he has elsewhere shown himself to be quite comfortable in publicly setting his achievements in the context of his divine support and lineage: consider, for example, the statuary in the Forum Augustum and the iconography on the Temple of Mars Ultor.

This leads us to what we might have expected all along, that astrology is for Augustus first and foremost an exercise in political posturing to fit the particular context. In certain contexts, it was politically expedient for the Emperor to fashion himself as an individual favoured and supported by the gods, and destined by the stars to achieve success. But when it comes to detailed cataloguing of his life's highlights, Augustus—much like his father in his historical biographical works[36]—denies any supernatural element the opportunity to undermine his own personal achievements.[37]

For the Emperor, then, astrology seems primarily to have been a piece of political craft that was both essential and dangerous to him, to be promoted in the right contexts and outlawed in the wrong ones, a complex that Tacitus would capture succinctly a century later in his reflection on the wildly optimistic predictions given by astrologers to the luckless Otho (Tac. *Hist.* 1.22):[38]

[35] For these and other influences on *Res Gestae*, see Cooley (2009) 30–41.

[36] See Chapter 4.

[37] This is not to say that *Res Gestae*, with its focus on earthly achievements and benefactions, might not have been designed to forward the case for his own official deification; see esp. Bosworth (1999).

[38] The text of Tacitus is taken from Wellesley (1989).

> urgentibus etiam mathematicis, dum novos motus et clarum Othoni annum observatione siderum adfirmant, genus hominum potentibus infidum, sperantibus fallax, **quod in civitate nostra et vetabitur semper et retinebitur**.
>
> Even the astrologers urged Otho into action, while they were confirming through their observation of the stars fresh revolution and a bright year for the man—this is a race of men treacherous to those in power, deceptive to those with aspirations, **a race which will always be both forbidden and maintained in our city**.

Augustan writers readily enter into the imperial spirit of this complex in their mixture of celebration of and discretion towards astrology. It is to these authors that we now turn.

6

Setting Gentlemanly Limits for Imperial Stellar Investigation in Vitruvius' *De Architectura*

Our first significant Augustan treatment of astrology occurs in a perhaps unlikely source, namely the architectural opus of Vitruvius. Everything we know about the author and the motivation behind his work comes from the detailed prefaces to each of its ten books. Vitruvius was evidently both a writer and a practising architect, and for this reason he is now generally believed to have been one of the *apparitores*, a group of state officials who helped the magistrates with their duties, thus providing continuity of expertise against a backdrop of yearly appointed politicians. Within this group, Vitruvius seems to have been both architect and scribe, and provides evidence of having been in the employ of Julius Caesar and Octavian/Augustus.[1] His major work, *De Architectura*, was probably composed during the early 30's to mid 20's BC and published in the mid-to-late 20's BC.[2] Its intended audience emerges gradually. First and foremost, it is dedicated to Augustus (*imperator Caesar*, 1 *praef.* 1), but Vitruvius subsequently widens the readership to include anyone who intends either to make practical use of the instruction or to increase his general learning: *non modo aedificantibus sed etiam omnibus sapientibus* (1.1.18).[3]

[1] For detail of Vitruvius' career, see esp. Anderson (1997) 40–4. For general acceptance of Vitruvius' position as *apparitor*, see Purcell (1983) 156; Rawson (1985) 86–7; Gros (1994); Masterton (2004); Nichols (2009) and, more cautiously, Anderson (1997) 42–3. For the intriguing but problematic theory that Vitruvius is the figure of Mamurra from Catullus' poetry, see e.g. Baldwin (1990) 430–1; Anderson (1997) 40–1.

[2] For the debate surrounding the date of the work, and whether or not it was circulated in serial or revised fashion, see Fleury (1990) xvi–xxiv; Baldwin (1990); Howe and Rowland (1999) 3–5.

[3] Cf. also 5 *praef.* 2–3, 5, 6.7.7. Vitruvius later makes the more grandiose claim that his work will be welcomed by 'all peoples' (*omnibus gentibus*, 6 *praef.* 7).

Recognition of such a wide-ranging audience for the work advises us against any simplistic assessment of its intended purpose.[4] We may start first with what Vitruvius himself suggests. The author offers three reasons for dedicating the work to Augustus (1 *praef.* 2–3): as a reciprocal gesture in the light of the favours the Emperor has given him;[5] as a celebration of the magnificent construction work that the Emperor has undertaken, and is continuing to undertake, in the city of Rome; as a means of enabling the Emperor to appreciate the technical skill involved in his construction work (which, of course, reflects back onto the author/architect).[6] For the more general learned reader who is seeking an acquaintance with, rather than active competence in, the art, the author boasts his novelty in the field (4 *praef.* 1), and pledges that his style—concise, well-ordered by topic and with technical terminology explained—is designed to accommodate the busy lifestyle of the literary dilettante (5 *praef.* 3). In constructing his readership in this way, Vitruvius appears to envisage *De Architectura* as an encyclopaedic reference work for the learned Roman in a similar vein to the later works of Strabo and Pliny the Elder.[7]

But caution must be exercised in taking at face value Vitruvius' authorial claims and the information he provides about Roman architectural practice and theory. There is now a general scholarly consensus that Vitruvius blurs the distinction between actual (and perhaps personal) practice and the projection of a more idealised picture of both architecture and its practitioner. This is perhaps no more keenly felt than in the extensive education that the author advocates for the architect (1.1.3), whom we are told must acquire a corpus of knowledge in a variety of disciplines: draftsmanship, geometry, history, philosophy, music, medicine, legal matters and astronomy (*astrologia*).[8] Although

[4] See most succinctly Fleury (1990) xxx–xxxvii.

[5] The nature of the 'favours' given to Vitruvius is obscure. The author speaks of *commoda* (1 *praef.* 2), which might mean a wage, financial bonuses or social privileges; see Fleury (1990) 61–2.

[6] Suetonius (*Aug.* 89.2) notes Augustus' enjoyment of technical treatises on architecture, when he suggests that the Emperor read to the Senate a work by Rutilius entitled *De Modo Aedificiorum*.

[7] See e.g. Callebat (1994) 31–4.

[8] Vitruvius' use of *corpus* suggests that all areas of his instruction come together to form an organic whole, thus reinforcing the impression that one needs to embrace all disciplines simultaneously; cf. 4. *praef.* 1; 5. *praef.* 5; 7 *praef.* 14, 9.8.15, 10.16.12; see further McEwen (2003).

all these disciplines relate to topics covered in the work, such wide-ranging learning (1.1.12 *encyclios . . . disciplina*) might seem optimistic. Perhaps Vitruvius is relying on a previous source (Pythius, 1.1.12) which harks back to a 'golden age' of the master architect who would carry out all aspects of the job himself. More persuasive, to my mind, is the view that this is a thoroughly modern touch from Vitruvius, an attempt to raise the intellectual profile of architecture in the minds of the elite: the encyclopaedic knowledge required for architecture has the effect of putting it on a par with the ideals of oratory, for which similar prescriptions are given, as well as the 'liberal arts' education of the elite polymath.[9]

As well as an intellectual agenda, there may also be a more personal and social dimension to Vitruvius' work. The figure of the architect often held an ambivalent position in the elite Roman mindset.[10] Cicero, for example, could praise the usefulness (*utilitas*) of architecture and the considerable learning (*prudentia maior*) required from its practitioners (*Off.* 1.151). This commendation is, however, only relative, as Cicero speaks about architects in a context of lower-grade occupations which take a wage and are, as such, unsuitable for the elite gentleman (*Off.* 1.150–1): it is noteworthy that he says that the vocation of architect is 'proper for those whose social position it becomes' (*eae sunt iis quorum ordini conveniunt honestae*). Consistent with this implicit snobbery, other ancient commentators attribute to architects the same characteristics that they give other low-class tradesmen, such as ignorance (e.g. Mart. 5.56.10–11). Moreover, in his specific role as *apparitor*, Vitruvius would have endured a curious intermediary status in Rome. His education and intimate connection with the ruling classes would have raised his profile; and yet his status as a wage-earner ranked him lower than equestrian class and associated him more readily with the working masses.[11] Against this background, the desire to ennoble the status of the architect and his craft might be a response to the author's own ambivalent position in society. Indeed, recent scholarship has effectively shown that, despite the reverential

[9] Vitruvius certainly demands more of the ideal architect than does Strabo, who calls only for an architect's competence in the fields of astronomy and geometry (1.1.13, 2.5.1). For the encyclopaedic education of the orator, cf. Cic. *De Or.* 3.126–7, 3.132–6; Quint. 1.10. Vitruvius may even be intimating that architecture is better than other elite arts, as it requires both intellectual rigour and practical experience; cf. 1.1.1–3, 1.1.11 *summum templum architecturae*.

[10] See in general Anderson (1997) 3–67, esp. 14–15.

[11] See Purcell (1983).

tone of its initial preface, Vitruvius is actually seeking to identify himself with the imperial elite, perhaps taking advantage of the changing status of technical experts under the emerging principate.[12]

It is Vitruvius' desire to explore and align himself with elite imperial sensibilities which is most interesting for our specific focus on his treatment of astrology. Having introduced the importance of astronomy for the architect (1.1.3, 10), Vitruvius duly proceeds, in his book dedicated to clocks and sun dials (Book 9), to map out the general constitution of the cosmos (9.1), followed by specific sections on the moon (9.2), the sun (9.3), and the constellations of the Northern and Southern hemispheres, respectively (9.4, 5). Only in 9.7–8—well over half the way into the Book—does the author turn to the specific topic of clock-making. The relevance of Vitruvius' extensive treatment of the stars in 9.1–6 has long been called into question.[13] Instead of a few principles based around the central indicators of time—the sun and the moon—Vitruvius has opted for a detailed exposition for its own sake.[14] For example, how does knowledge of the northern and southern constellations (9.4, 5), and of the orbiting times of some of the more remote planets (9.1.6–10), aid the clock-maker? Vitruvius does not tell us, and in fact it is a notable trait of the exposition in 9.1–6 that the author never breaks off to point out the practical consequences of the information he relays.

Despite the author's claims, therefore, that information will be provided for practical usage (9 *praef.* 18), the detail in 9.1–6 seems to be geared more towards the ongoing elevation of architecture as a learned, wide-ranging discipline and towards the general learned reader, thus providing further evidence of the intellectual interest in stars at the time of writing. This section—which might be aptly numbered among

[12] Vitruvius seeks to align himself with the elite and its activities by playing down the aspect of payment for his craft (Masterton 2004) and by joining in the criticism of those *apparitores* who are financially ambitious (Nichols 2009). König (2009) offers a bolder reading: Vitruvius effectively elevates himself above the level of the Emperor by suggesting that mastery of architecture is ultimately borne from the architect's genius, which is both innate and insusceptible to instruction; for the 'enlarged' status of the architect in Vitruvius, see also Cuomo (2011). For the emergence of technical experts under Augustus, see Wallace-Hadrill (1997).

[13] See esp. Soubiran (1969) xv–xvii, xxxiv–xxxviii.

[14] It should be noted that even the sections on the moon and sun are of little practical use, as they have nothing to say on the variation in length of days and hours, the importance of which for clock-making had been identified at 9.1.1; see Soubiran (1969) 130, 142.

Vitruvius' learned 'digressions'[15]—is the work of an intellectual dilettante writing for fellow enthusiasts. The occasional contradictions in the Vitruvian cosmos point to an author who is using a range of sources, with little attention to internal inconsistencies.[16] Moreover, the omissions in and simplifications of Vitruvius' heavenly system indicate a desire to present a pleasing narrative which is easy to follow and retains attention. Indeed, our author appears to be motivated by a desire not so much for accuracy and completeness as for avoidance of causing offence through long-windedness: cf. 9.7.7 *quas ob res non pigritia deterritus praetermisi, sed ne multa scribendo offendam*. True to this aim, Vitruvius is also keen to provide more accessible analogies to help the reader visualise the cosmic system he is describing: the wheels around the polar hinges work 'as on a lathe' (9.1.2 *uti in torno*); planets travel west to east across a circuit of greater magnitude than the constellations, 'just as if they were at different points on a staircase' (9.1.5 *ut per graduum ascensionem percurrentes*); the apparently differing speed of planets, due to different-sized orbits, may be usefully compared to ants on differing grooves of a potter's wheel (9.1.15).[17]

Vitruvius' enthusiasm for stellar exploration does not, however, extend to the topic of astrology. In 9.6.2–3, our author turns to the use of stars to predict the future—in both its genethliological and astrometeorological forms—in the following manner:[18]

> [2] ceterum ex astrologia quos effectus habeant signa XII, stellae V, sol, luna ad humanae vitae rationem, **Chaldaeorum ratiocinationibus est concedendum**, quod propria est eorum genethlialogiae ratio, uti possint ante facta et futura **ex ratiocinationibus** astrorum explicare. eorum autem inventiones reliquerunt, in quae **sollertia acuminibusque** fuerunt **magnis**, qui ab ipsa natione Chaldaeorum profluxerunt ; primusque Berosus in insula et civitate Coo consedit, ibique aperuit disciplinam, post ei studens Antipater, iterumque Achinapolus, qui etiam non e nascentia sed ex conceptione genethlialogiae rationes explicatas reliquit.

[15] For others, cf. e.g. 2.1.1–3 (history of humanity), 2.2 (the four elements), 6.1 (ethnography), 7.8 (properties of mercury).

[16] For the shortcomings of Vitruvius' cosmos, see Soubiran (1969) xxxviii–xlvii; Rawson (1985) 165–6.

[17] Two of these analogies were clearly influenced from Vitruvius' sources. For the analogy from craftmen's tools, cf. Arat. 529–33, Apul. *Mund.* 1; for the ants on the potter's wheel, cf. Cleomedes, *De motu circulari* 1.3.16. For the contradiction in Vitruvian thought caused by this analogy, however, see Soubiran (1969) xliii; Rawson (1985) 165–6.

[18] The text of Vitruvius is taken from Soubiran (1969).

[3] de naturalibus autem rebus Thales Milesius, Anaxagoras Clazomenius, Pythagoras Samius, Xenophanes Colophonius, Democritus Abderites rationes, quibus e rebus natura rerum gubernare[n]tur, quemadmodumcumque effectus habeat, excogitatas reliquerunt. quorum inventa secuti siderum ortus et occasus tempestatumque significatus Eudoxus, Euctemon, Callippus, Meto, Philippus, Hipparchus, Aratus ceterique ex astrologia parapegmatorum disciplinis invenerunt, et eas posteris explicatas reliquerunt. **quorum scientiae sunt hominibus suspiciendae**, quod **tanta cura** fuerunt ut etiam videantur **divina mente** tempestatium significatus post futuros ante pronuntiare. quas ob res **haec eorum curis studiisque sunt concedenda.**

[2] But as to the other branch of stellar activity—what effects the twelve zodiacal signs, the five planets, the sun and the moon have towards the reasoning of human life—**this branch of activity must be conceded to the reasoning powers of the Chaldeans**, because the reasoning behind their genethliological astrology, the ability to explain the past and the future through **reasoned examination** of the stars, is their own. But those who flowed forth from the very nation of the Chaldeans have bequeathed to us their discoveries in this area, discoveries which expended **ingenuity** and **great sharpness of mind**. First there was Berosus who settled on the island and state of Cos and there opened up the discipline; there followed his student Antipater, and after him Achinapolus, who even bequeathed to us theories of horoscopes calculated not from the moment of birth but from the moment of conception.

[3] But in the realm of natural science, it was Thales of Miletus, Anaxagoras of Clazomenae, Pythagoras of Samos, Xenophanes of Colophon and Democritus of Abderitus who bequeathed to us the theories they generated relating to the things by which the natural world is governed and in what manner it has its effects. Eudoxus, Euctemon, Callippus, Meto, Philippus, Hipparchus, Aratus and others followed the discoveries of these men and found out from the discipline of parapegmata (itself derived from astrology) the rising and setting of the stars and the indications of weather: and they left these things explained to posterity. **The science of these men must be held up in admiration by humanity**, because it was achieved with **such great care** that they even seemed to have a **divine mind** when they were announcing in advance indications of weather which was to come to pass later. On account of these matters, **this science must be conceded to the care and diligence of these men.**

Vitruvius' brevity here stands in marked contrast to his expansive treatment of the stars in 9.1–5. Whereas astronomical observation was evidently felt to be common intellectual property, astrology is conceded to experts who are distinctly foreign: horoscopic astrology is the realm of the Chaldeans and Babylonians (Berosus), whereas astrometeorology is the remit of the Greeks.[19] Vitruvius offers consistent respect and praise for astrologers and their craft: their science is one of reason and knowledge (*ratiocinationibus, scientiae*) and their wisdom and intellect are considerable (*sollertia acuminibusque . . . magnis*; cf. also *tanta cura* and *curis studiisque* used of those who predict the weather); more grandly, those who can predict the future—in this case weather—are felt to possess a 'divine mind' (*divina mente*).[20] But the exclusivity of the science is just as pronounced as any respect for it: the sustained use of gerundives in this section (*concedendum, suspiciendae, concedenda*) rams home the central message that this particular branch of stellar activity, the reading of stars to predict the future, should remain exclusive to these hallowed foreign experts.

It is not our author's intention to withhold information per se which is remarkable: Vitruvius does, for example, make a point of omitting instruction on ladders and cranes for military purposes (10.16.1) and on machinery for defence (10.16.2). But on these occasions, he is doing so either out of respect for the reader—the construction of ladders is the work of a soldier—or because strategies for defence vary according to the nature of the particular enemy.[21] The remarkable point about 9.6.2–3 is that Vitruvius' discretion is actually deference: the reader simply does not have the intellect, wisdom and divinity of mind to grasp such a science, and so must yield to specialist practitioners.

Taking an overview of Book 9, it is evident that Vitruvius has strayed somewhat from what we might call customary architectural instruction. First, the very inclusion of clocks and sundials in the work may go beyond the requirements of a study of architecture, in the light of the fact that no other ancient author appears to include it within the discipline.[22] Secondly, having included clocks and sundials, our author

[19] It is important to note that the foreignness of technical knowledge does not elsewhere present a barrier to Vitruvius' handling of the topic; see esp. 5.4.1, in which Vitruvius tackles the Greek theory of musical harmony, despite his admission that it is both obscure and difficult (*obscura et difficilis*).

[20] The use of *divina mens* puts the Chaldeans in the highest company. The phrase is elsewhere used to describe the Emperor (1 *praef.* 1), the world power of Rome (6.1.11) and the workings of the universe (9.1.1, 9.5.4), in what effectively amounts to the earliest articulation of the 'cosmos and imperium' conceit.

[21] See further Cuomo (2011) 313–14.

[22] See McEwen (2003) 230. McEwen (2003) 229–50 argues that the entire section on clocks and time-keeping is a response to different aspects of emerging Augustan discourse, namely the connection between ruler and the sun, and control over time in the form of reformation of the calendar and transposition of Egyptian obelisks. But McEwen offers no discussion of Vitruvius' particular stellar priorities in Book 9.

goes beyond the call of his instructional duty by including stellar material which has no bearing on the practicalities of clock-making, particularly the description of constellations in the cosmos and the brief treatment of astrology, which serves principally to advertise both the credibility and the exclusivity of the science.

The admiration for astrology expressed here stands in marked contrast to the pejorative overtones that have previously marked Roman discussion of the Chaldeans. Soubiran notes the stark difference in tone on astrology between the reverential Vitruvius and the combative Cicero of *De Divinatione*, and accounts for this in terms of Vitruvius' beliefs and his general penchant for all things oriental.[23] But it seems more fitting to think in terms of a change in political emphasis for astrology that has occurred between the times of the two authors. To my mind, we may regard Vitruvius as the first extant Augustan author to attempt to set out for the learned Roman reader what he sees as acceptable limits of stellar enquiry in the new imperial age. First, a detailed description of the constellations, which rehearses the well-known text of Aratus, provides an appropriate celebration of the stellar underpinning to astrology.[24] Celebration of astrology is also achieved, more subtly and indirectly, by acknowledgement of Julius Caesar's deification through catasterism (1 *praef.* 2).[25] But Vitruvius advises his readers against venturing further: the influence of the stars on different aspects of human life is acknowledged, but its intricacies are respectfully deferred and given a distinct air of exclusive authority. Vitruvius ends Book 9 with the following remark (9.8.15):

> quae sunt in horologiorum descriptionibus rationes et apparatus, ut sint ad usum expeditiores, **quam apertissime potui** perscripsi.
>
> About the calculations and instruments which are involved in the arrangement of dials, I have now written in full, **as openly as I have been able**, so that they might be more readily applied to use.

Granted that *quam apertissime potui* is not uncommon as a closing editorial remark from our author (cf. 3.5.15, 5.4.1, 6.8.10, 10.6.4), it is tempting to give this phrase a particularly pointed force here, as the author both reflects on his ability to explain difficult points of technical detail and broadcasts his awareness of appropriate levels of disclosure.

23 See Soubiran (1969) liv–lv and 202 ('la crédulité de Vitruve').

24 For Vitruvius' use of Aratus and Eudoxus, see Soubiran (1969) 143–99 *passim*.

25 For catasterism as a branch of astrology, see Chapter 3.

It is perhaps natural for Vitruvius, a man striving for elite recognition, to make his statements of intent with regard to astrology in such a direct, even didactic manner. All forms of predictive astrology remain the preserve of a select number of foreign experts, outside the reach of Vitruvius' general (i.e. non-imperial) reader, whose knowledge is not to be divulged. For this reader, Vitruvius offers his own model of appropriate stellar enquiry through a meditation on Aratus, whose poem celebrates and endorses the cosmic backdrop for astrology without disclosing any of its secret workings. By reflecting upon appropriate boundaries for stellar discourse, the author seeks to align himself with the political strategies of his imperial patron. And we will see that the spirit of Vitruvius' meditation in this area is felt, and renegotiated, in stellar literature throughout Augustus' reign.

7

Concession, Abstinence and Abortion

Horace, Virgil, Hyginus and Ovid

Vitruvius, then, sets up different levels of disclosure with regard to stellar knowledge, with one of the highest levels of technical knowledge, genethliological astrology, respectfully conceded to experts. Such a discourse of concession and abstinence finds dynamic expression in a variety of contemporary authors as they engage with the controversial politics of the science.

CONCESSIONS FROM THE NATION'S FATHER: VIRGIL'S *AENEID*

Even through the medium of a mythological epic Virgil finds means of bringing his own subtle brand of celebration and discretion to bear on astrology. Respected seers in the epic are given power over the stars, a projection of contemporary divinatory expertise back into the mythical past. So the Trojan Helenus and the Etruscan Asilas are endowed with the powers of astrology before their time (Verg. *A.* 3.359–61, 10.175–7):[1]

> Troiugena, interpres divum, qui numina Phoebi,
> qui tripodas, Clarii et laurus, **qui sidera sentis**
> et volucrum linguas et praepetis omina pennae . . .
>
> O son of Troy, interpreter of the gods, you who feel the will of Apollo, the tripods and laurel of the Clarian, **you who feel the stars**, the tongues of birds and the omens of the flying wing . . .

[1] See Horsfall (2006) 278. The text of Virgil is taken from Mynors (1969).

tertius ille hominum divumque interpres Asilas,
cui pecudum fibrae, **caeli cui sidera parent**
et linguae volucrum et praesagi fulminis ignes . . .

Third comes Asilas, that interpreter of gods and men, whom the entrails of cattle, **whom the stars of heaven obey**, as well as the tongues of birds and the fires of presaging thunder . . .

Furthermore, the predictive power of the stars may be acknowledged in some passing phrases. Consider Virgil's description of Dido's preparations for death (Verg. *A.* 4.517–20):

ipsa mola manibusque piis altaria iuxta
unum exuta pedem vinclis, in veste recincta,
testator moritura deos et **conscia fati**
sidera.

She herself, with holy meal and holy hands, beside the altar, with one foot bereft of footwear and with clothes torn, gives evidence that she will die to the gods and **the stars that have knowledge about her fate**.

Conscia fati sidera may be rendered tamely as if the stars were merely 'witnesses' to Dido's doom, but the exact same phrase exerts a strong astrological charge in the opening lines of Manilius (1.1–2), and Servius may well be right to insist on a similar astrological reading here.[2]

But if Virgil implicitly lends endorsement to the science of astrology, he directs Romans away from actually practising it via the most famous dictates of Anchises (Verg. *A.* 6.847–53):

'excudent alii spirantia mollius aera
(credo equidem), vivos ducent de marmore vultus,
orabunt causas melius, **caelique meatus**
describent radio et surgentia sidera dicent:
tu regere imperio populos, Romane, memento
(hae tibi erunt artes), pacique imponere morem,
parcere subiectis et debellare superbos.'

'Others—for I well believe it—will hammer out the breathing bronze more softly and draw forth a lifelike face from marble; they will plead cases more competently, **and they will trace with the rod the paths of the sky and will speak about the rising stars**. You, Roman, remember to rule the peoples with power—these will

[2] Cf. Serv. *Aen.* 4.519 *conscia fati sidera: id est planetas, in quibus fatorum ratio continetur. conscia sidera* reappears at Verg. *A.* 9.429. See further DeWitt (1930) 22.

> be your arts—and to add to peace a settled way of life, to spare the oppressed and to war down the proud.'

While 6.849–50 refer most readily to astronomy, astrology is also suggested by *surgentia sidera dicent*—in view of the fact that horoscopes take special note of rising constellations—and Servius once again encourages us to read Virgil against this astrological background with his reference to the Chaldeans.[3] But there is no sense of contempt in Anchises' words here: in keeping with the approach of Vitruvius, foreign arts prominent in Augustus' own age are respectfully conceded to others, while the Romans are charged with pursuing a different set of activities.[4]

PRACTISING ABSTINENCE: HORACE

Horace *Carm.* 1.11 engages in a more direct manner with the practice of astrology:[5]

> tu ne quaesieris, scire nefas, quem mihi, quem tibi
> finem di dederint, Leuconoe, nec Babylonios
> temptaris numeros. ut melius, quidquid erit, pati,
> seu pluris hiemes seu tribuit Iuppiter ultimam,
> quae nunc oppositis debilitat pumicibus mare
> Tyrrhenum! sapias, vina liques et spatio brevi
> spem longam reseces. dum loquimur, fugerit invida
> aetas. carpe diem, quam minimum credula postero.

> Do not enquire, Leuconoe—it is unlawful to know—what end the gods have granted for me and for you, and do not try out for yourself the calculations of the Babylonians. How much better it is to endure whatever happens, whether Jupiter has allotted us additional winters, or whether this one is our last, which now wears out the Tuscan sea upon the opposing rocks! Be wise, strain the wine, and since life is short cut back far-reaching hopes. While we speak, envious time will already have fled. Pluck the day, putting as little trust as possible in tomorrow.

[3] Cf. Serv. *Aen.* 6.848 . . . *per astronomiam Aegyptios et Chaldaeos* (*indicat*).

[4] See Austin (1977) 261: 'Virgil looks at the Greek genius with respect and honour, and without jealousy'; Williams (1964) 60–1.

[5] The text of Horace is taken from Shackleton Bailey (1985).

Most famous for its final sentiment, Horace opens this poem (1–3) with a direct warning to Leuconoe not to engage in horoscopic astrological enquiry, the type of astrology which would be able to determine the length of a person's life (*finem*) and would require mathematical calculations derived from foreign experts (*Babylonios . . . numeros*). In the light of the rest of the poem, in which Horace exhorts his girlfriend to live for the moment, it would be fair to argue that the poet is advocating the view that horoscopic astrological knowledge is unprofitable. But it would be wrong to use this poem, as some scholars have, as evidence of a general disregard of or scepticism towards astrology on Horace's part.[6] In fact, Horace specifically endorses horoscopic astrology at *Ep.* 2.2.183–9, when he suggests that mortal life is governed by the interactions between the natal star (187 *natale . . . astrum*) and the human's own divine spirit, the *Genius*.[7]

Coming back to *Carm.* 1.11, it is clear that Horace is dealing only with one specific branch of astrology, and that he is not motivated by scepticism. Rather, his specific objection to astrological consultation of this kind is that it amounts to *nefas* (1).[8] *Nefas* can certainly refer to a violation against natural or divine order, but one wonders whether it also points to a more earthly, legal sort of violation.[9]

Though it is difficult to date the *Odes*, there is general scholarly consensus that the earliest poems may date from the very late 30's BC, and that Book 1, possibly along with Books 2 and 3, was published in the mid-to-late 20's BC.[10] It is more than likely, then, that Horace's poem was both written and received against the political backdrop of Octavian's recent high-profile culling of astrologers

[6] For such a view, see e.g. Syndikus (1972) I.460; Rudd (1989) 145; Lyne (1995) 119–20.

[7] On interpreting the combination of Roman and foreign elements in Horace's fatal system here, see discussion of *Carm.* 2.17 later.

[8] In a poem addressed to Maecenas, Horace uses similar language in dissuading mortals from seeking information about the future (albeit without specific reference to astrology); cf. *Carm.* 3.29.29–32 *prudens futuri temporis exitum/caliginosa nocte premit deus/ridetque si mortalis* ***ultra/fas*** *trepidat* ('a wise god hides in the darkness of night the outcome of future time, and he laughs if a mortal agitates **beyond what is lawful**').

[9] For *nefas* in the more generic sense of 'wrong', cf. e.g. Verg. *G.* 1.505; Hor. *Epod.* 5.87; *Carm.* 1.18.10; Ov. *Met.* 6.613, *OLD* s.v. 2c.

[10] See in general Nisbet and Hubbard (1970) xxvii–xxxviii; Hutchinson (2002). From the potentially datable elements of the poems in the first Book of *Odes*, Hutchinson (2002: 522–3) establishes parameters of 30–25 BC. The date of publication for this book may range from 26 BC (Hutchinson 2002: 528) to 23 BC (Nisbet and Hubbard 1970: xxxvii).

from Rome in 33 BC. Against this political background, Horace's injunction may serve to steer his girlfriend away from activity that might be deemed politically inappropriate in the mortal Roman world. In this vein, the two verbs used in the injunction—*quaesieris* and *temptaris*—might constitute a double warning to Leuconoe: she is advised to avoid both 'seeking' astrologers and 'trying out' their calculations for herself.[11]

The central thrust of the poem—to live for today—may well be a commonplace in ancient thought.[12] But Horace has given his poem contemporary political resonance and has, in the process, advertised himself as a lawful upholder of the imperial position on horoscopic astrology. I should add at this juncture that I do not feel that Horace's alignment with imperial sensibilities in *Carm.* 1.11 is in any way compromised by the horoscopic detail found in *Carm.* 2.17, a much discussed topic to which we should turn briefly.

In the middle of *Carm.* 2.17, a poem that offers a charming expression of loyalty from Horace to Maecenas, the poet speaks, in astrological terms, about their apparently shared fate, making reference to both signs of the zodiac and (deified) planets (17–25):

> seu Libra seu me Scorpios aspicit
> formidolosus, pars violentior
> natalis horae, seu tyrannus
> Hesperiae Capricornus undae,
>
> utrumque nostrum incredibili modo
> consentit astrum. te Iovis impio
> tutela Saturno refulgens
> eripuit volucrisque Fati
>
> tardavit alas, . . .

> Whether the Balance or the dreaded Scorpion looks towards me as the most potent part of my natal hour, or Capricorn, the lord of the Hesperian wave, the star for both of us is miraculously in agreement. The shining guardianship of Jupiter snatched you away from impious Saturn and slowed down the wings of Fate in flight . . .

[11] For *quaero* and its cognates as the appropriate verbs to describe consultation of astrologers, cf. Prop. 2.27.1–2; Ov. *Met.* 1.148; Tac. *Ann.* 3.22.2. For Roman women trying out astrology for themselves, see the disparaging comments of Juvenal (6.553–91, esp. 572–81).

[12] See esp. Nisbet and Hubbard (1970) 138–9.

There is no doubting Horace's proficiency in the technical terminology of astrology in this section.[13] But can we go further and maintain that Horace is disclosing, albeit in shorthand form, a real horoscope that had been cast for him (and his patron)? While this view has its supporters,[14] I find myself in agreement with those scholars who advise caution in this regard. First, the repetition of *seu* (17, 19) suggests a more speculative commentary from Horace, as the poet, unaware of the specificities of his own horoscope, is ranging through a list of possibilities.[15] Secondly, the constellations mentioned exert such a conspicuous imperial charge as to suggest that astrology is here being employed for politically ideological reasons.[16] In particular, as we have seen, Libra, and especially Capricorn, have associations with Augustus: is Horace intimating that both Maecenas and himself are under the same stars that protect the Emperor or even, by extension, that both are under the protection of the Emperor?[17]

A more generalised interpretation of astrological material in these lines is also supported by the wider context of the poem, where we discover that the stars are just one of a number of possible guardian influences on mortal life: indeed, in the case of the poet's recent accident, it is Faunus who is assigned the role of guardian influence (27–30).[18] The presence of Faunus does not, as some scholars suggest, automatically undermine or make fun of the previous engagement with horoscopic astrology.[19] I am not denying the poet a gentle touch of humour in this section, as he contrasts his own guardian, the woodland deity Faunus (28), with Jupiter (22–3), the mighty guardian of his esteemed patron.[20]

[13] See esp. Kidd (1982) 88–93; Bollók (1993) 14–16.

[14] See Bouché-Leclercq (1899) 551; West (1991); Bollók (1993).

[15] See McDermott (1982) 225; see also Bradshaw (2002: 13), who argues that the three specific signs mentioned in 17–20 may well be meant to act as an exemplary selection, a strategy which is common in Horace.

[16] For this (more popular) interpretation of the astrology in this section, see e.g. Dicks (1963: 70–2) and esp. Kidd (1982: 88–90), who also remarks upon the literary homage being paid to Aratus in these lines. For a different take on the ideological significance of these signs, see Coutts (1964).

[17] See Dicks (1963) 71; Bradshaw (2002) 13. For Augustus' association with Libra and especially Capricorn, see Chapter 5.

[18] The choice of Faunus is intriguing, and seems to be connected with the equally intriguing use of *Mercurialium* (29). For discussion of the possibilities, see Nisbet and Hubbard (1978) 286; Lyne (1995) 120.

[19] For this view, see e.g. Nisbet and Hubbard (1978: 273–4) and Lyne (1995: 119–20), who both argue that Horace is poking fun at his patron's serious interest in astrology.

[20] For the humour of this poem, see esp. McDermott (1982).

But the juxtaposition of different forms of guardian/supernatural influence— drawn respectively from the foreign science of astrology and the indigenous, Italian realm of native religion—might operate as a respectful commentary on contemporary Rome, in which Roman religion and astrology stand side by side in the influencing of a human's life: after all, both astrology and traditional Roman religion are publicly championed by the Emperor. It is in this spirit of celebration that I suggest we also take the combination of foreign and Roman supernatural forces in *Ep.* 2.2.183–9 (natal star and Genius: see earlier discussion).

As with *Carm.* 1.11, Horace in *Carm.* 2.17 takes traditional poetic motifs—such as the idea of lives bound together[21]—and gives them contemporary Roman relevance by engaging with astrology in a politically sensitive manner. Horace celebrates the power of natal astrology in a general way, which does not amount to specific disclosure of a real horoscope. In the process, he pays homage to the Emperor both directly, by drawing attention to the signs important to his own (publicised) horoscope, and indirectly, by bearing witness to the harmony that now exists in Rome between the foreign art of astrology and the traditional Roman religion of native Italy.[22] The political thrust of this poem may be particularly relevant if its dramatic date can be fixed to the period 26–24 BC, against a backdrop of Augustus' renovation and building of public structures which celebrate both traditional religion and astrology (in the form of the Pantheon).[23]

HYGINUS AND OVID: DISCOURSES OF ABORTION?

Further to the motif of astrological abstinence, it would be worth turning briefly to the prose treatise on the stars which is typically referred to as *De Astronomia*. It is attributed in the manuscripts to an individual named Hyginus, and dedicated to M. Fabius. There is general scholarly consensus that this is the work of C. Iulius Hyginus (*c.* 64 BC—AD 17), the freedman of Augustus from Spain, whom the

[21] See Nisbet and Hubbard (1978) 272.

[22] See also Herbert-Brown (2002: 116), who speaks of Horace and Propertius as 'reflecting contemporary eclecticism in religion without privileging any one system over another'.

[23] Hutchinson (2002: 524–5) suggests a dramatic dating of 25–24 BC for Horace's second book. For the potentially astrological overtones to Augustus' Pantheon, see Chapter 5.

Emperor famously gave charge of the library on the Palatine (Suet. *Gramm.* 20), and to whom the sources attribute several other works of literature. The dedicatee might well be Paullus Fabius Maximus. Based upon these identifications, the work has been tentatively dated to the period 11–3 BC.[24] The work has a discernible four-part structure, which in modern editions is denoted by books. Following a preface, it opens with a general discussion of the cosmos and the celestial sphere (Book 1), before moving on to what appears to be the heart of the work, a discussion of no fewer than 42 catasterisms (Book 2). The same constellations are then handled again, but this time with regard to their position in the sky, relative to other constellations, and the number of stars that make them up (Book 3). The treatise concludes with a discussion of the movement of the constellations and the planets across the heavens (Book 4).

The treatise is heavily influenced by Aratus—whose apparent obscurity it is Hyginus' expressed intention to elucidate (*praef.* 67–9)[25]—and the longest section, on catasterisms, seems to be particularly indebted to the now fragmentary *Catasterismoi* of Eratosthenes.[26] As well as a source of general interest to the learned reader (*praef.* 5–6), it has been suggested that Hyginus' work may also have served the practical function of enabling readers to construct their own model of the *sphaera*.[27]

For our purposes, it is noticeable that Hyginus' stellar investigation follows in a similar cautionary manner to that of Vitruvius. It is first and foremost an intellectual treatise that makes use of the astronomical texts of Hellenistic predecessors. However, even though there is very little discernibly Roman, and nothing discernibly imperial, in the work,[28] Hyginus pays lip service to his imperial benefactor by showing

[24] For the debate concerning authorship and date, see esp. Le Boeuffle (1965) 286–94, (1983a) xxxi–xli. The identification of the author with the Augustan Hyginus is certainly accepted by the most recent Teubner edition of the text; see Viré (1992) iii–v.

[25] The text of Hyginus is taken from Viré (1992). Hyginus offers further specific criticism of Aratus at 2.2 and 4.1.

[26] Hyginus mentions Eratosthenes no fewer than 21 times in his work, 19 of those in the section on catasterisms; Aratus is mentioned on 15 occasions and Callimachus on seven occasions [figures are taken from Chiabò and Roberti (1990)].

[27] For the popularity of the model *sphaera* in the 1st century BC, cf. Cic. *Rep.* 1.14; *ND* 2.88; see further Le Boeuffle (1983a) ix–xii; Evans and Berggren (2006) 27–34.

[28] The only Roman author to whom Hyginus makes direct reference is Cicero, whose version of Aratus' poem is referred to at 3.29 and 4.3. It is possible, however, that the work shows additional influence from Catullus and Nigidius Figulus; see further Le Boeuffle (1983a) xvi–xviii. It is intriguing, however, that Hyginus does not take the opportunity to incorporate imperial catasterism into his project.

an awareness of his stellar sensibilities. In the preface, Hyginus pledges a discussion of the signs of the zodiac 'as well as their influences', a move which anticipates subject matter of an astrological flavour.[29] But when our author turns properly to the topic of zodiacal signs at 1.8 and 4.5, in the context of their role during seasonal change, the link is not causal, as Hyginus speaks merely of the sun *indicating* seasons by coursing through the constellations.[30] He concludes 4.5 in similar vein:

> priores enim astrologi, cum omnes res ad duodecim partes revocarent ut menses et horas et latitudinem signorum, itaque et signa, **per quae res omnes significantur**, duodecim voluerunt esse.

> For the ancient astronomers, since they were reducing all things to twelve parts—such as the months, hours, and latitude of the signs (i.e. 12°)—in the same way they wanted there to be twelve signs of the zodiac, **through which all things are signified**.

By the end of the work, Hyginus has left a mixed impression: do the stars influence or merely indicate? One could argue that our author has raised the general issue of stellar influence at the beginning of his work only to resort later to the safe topic of astrometeorology—far removed from the realm of astrological influence on mortals—when he comes to more detailed discussion.[31]

A more interesting example of astrological discretion can be found in Ovid's *Fasti*, as argued in a fascinating piece by Herbert-Brown (2002). Centring the discussion on the eulogy to stargazers (*Fast.* 1.295–310), Herbert-Brown argues persuasively that Ovid is including astrologers within this hallowed group, individuals who are far removed from the stellar naïvety of primitive leaders and who are, in the case of imperial court astrologers, far removed from the aspirations and career paths of the contemporary Roman senator (1.297–304). Inspired by these experts, Ovid proposes to tackle astrology for himself within the poem (1.305–10):[32]

> admovere oculis distantia sidera nostris
>   aetheraque ingenio subposuere suo.
> sic petitur caelum, non ut ferat Ossan Olympus
>   summaque Peliacus sidera tangat apex.

[29] See *praef.* 31–2 *exinde zodiacum circulum definivimus* ***et eius effectus***.
[30] Cf. e.g. 1.8 *ostendit, significat*; 4.5 *demonstratur*.
[31] For astrometeorology as a safe topic for imperial writers, see Chapter 8.
[32] The text of Ovid's *Fasti* is taken from Alton, Wormell and Courtney (1988), except for the reading *nostris* for *mentis* in *Fast.* 1.305 (for which, see Green 2004: 141–2).

> nos quoque sub ducibus caelum metabimur illis,
> ponemusque suos ad **vaga signa** dies.

> They moved the distant stars to our own eyes, in that they subjected the heavens to their own intellect. This is the way to seek heaven, not by means of piling Ossa on Olympus and having the summit of Pelion touch the highest stars. I also, under the leadership of those men, will measure out the sky and will assign to the **planets** their own days.

From an imperial standpoint, Ovid's enthusiasm for astrology reaches potentially worrying heights in the final two lines, as the poet proposes to become a practitioner of astrology by giving due attention to the precise dating of planets, unmistakeably captured by *vaga signa*.[33] But any such concern turns out to be unfounded, as the extant poem offers no discernible astrological discourse and no discussion of planets. Herbert-Brown argues imaginatively that this may be the result of a late extraction of astrological material from the poem, possibly after Augustus' edict of AD 11, as Ovid reassessed its sensitivity from exile; on this reading, *vaga signa* in 1.310, overlooked by the poet in his revision, represents the only clue that excision of this sort has occurred.[34] This is to overstretch the case here, to my mind, as any such reading requires us both to accept significant interpretative baggage behind *vaga signa* and to balk at the sheer clumsiness of a poet who would have overlooked his own initial eulogy to stargazers in any revision of stellar material. Nevertheless, in the context of our discussion, Herbert-Brown's valuable observations do point, perhaps more plausibly, to a recognised literary game of astrological abortion, games such as those played by his contemporaries Germanicus and Manilius, which involve the allurement of a planetary discussion that is perpetually withheld.[35]

[33] For the use of *vagus* to indicate the planets, etymologically connected to πλάνητες ἀστέρες ('the wandering stars'); cf. e.g. Cic. *ND* 2.80 *vagas stellas*; Man. 1.52 *sideribus . . . vagis*; 2.742–3 *sidera . . . vaga*; Luc. 9.12 *stellasque vagas*. The proposed astrological thrust to Ovid's stellar discussion may also render more imperially confrontational the poet's opening to the eulogy at 1.295–6: ***quid vetat*** *et stellas . . . dicere?* What may have initially read as a rhetorical generic enquiry into the appropriateness of including stars in a calendrical poem starts to read, with hindsight, like a politically defiant stance towards an Emperor keen to curb the dissemination of astrological enquiry; see Herbert-Brown (2002) 122.

[34] See Herbert-Brown (2002) 121–8. For Ovid's revision of *Fasti* from exile, see Green (2004) 15–25.

[35] For Germanicus, see Chapter 8.

8

Stars and Storms: The Development of Stellar Causation in Astrometeorology

Astrometeorology, the science which uses the motion of stars as a means of predicting weather patterns and seasonal progression on earth, is the oldest recorded practical function of stargazing. It goes back at least as far as Hesiod in the 8th century BC, and remained probably the most respected branch of astronomy/astrology across the centuries.[1] We see in Hesiod a rather basic system—a few rules involving a few stars for the benefit of farmers and sailors—but this developed into a more wide-ranging system by the 3rd century BC, which coincided with the emergence of the *parapegma*, an instrument that helped farmers track the different celestial phenomena by means of a peg that could be moved from day to day.[2]

THE CURIOUS CASE OF THE DOG STAR IN THE SUMMER TIME: ASTROMETEOROLOGY BEFORE THE 1ST CENTURY BC

For my purposes, the most important question concerns causation: what role were the stars thought to play in the creation of atmospheric

[1] Astrometeorology, certainly in its original form, is a science which slips between our modern conceptions of 'astronomy' and 'astrology'. In order to achieve greatest clarity in this discussion, I refer to as 'astronomical' occasions on which the stars are evoked merely as signs of weather, and 'astrological' occasions on which the stars are deemed to be causes of weather. For the ancient respect for astrometeorology, see most notably the endorsement by Sextus Empiricus, who exempts this branch of star-reading from his general attack on the subject (*Adv. Math.* 5.1–2).

[2] See in particular Lehoux (2007).

conditions on earth? Our most ancient poetic sources, Homer and (especially) Hesiod, provide a strong theological import, in that the stars are conceptualised as *signs* of forthcoming weather, passive instruments of instruction sent by powerful deities, especially Zeus, who control the weather.[3] This approach is modified in the most influential poetic treatment of the stars, the *Phaenomena* of Aratus (3rd century BC), whose predominant view is that a benevolent Zeus has set up the stars, along with other natural phenomena, to act as trustworthy signs of forthcoming weather.[4] Aratus has very little to say about what connection there might be between the sign and the signified, and there is certainly no straightforward indication anywhere in the poem that the stars actually cause the phenomena with which they coincide temporally[5] —that is, with one curious exception for which all three of the Greek poets mentioned earlier appear to make room.

From Homer onwards, the Dog Star is popularly interpreted as if it were an independent agent and is charged with directly causing the hot weather that accompanies its rising in summer.[6] In fact, the popular name for the star, Sirius, is often felt to derive directly from its heat-bringing properties.[7]

What are we to make of this? I am not aware of any scholar who treats the Dog Star as an exception in Greek astrometeorological thinking. On the contrary, the general scholarly position is to treat the Dog Star as emblematic of a widely held and seemingly timeless view

[3] In Homer, Zeus is sometimes referred to as 'cloud-gatherer' (*Il.* 1.511, *Od.* 5.21) and is believed to cause thunder (*Od.* 12.415), lightning (*Il.* 8.133–4) and rainbows (*Il.* 11.27–8). But other gods such as Hera, Athene, Poseidon and Kalypso can also initiate weather in the Homeric tradition. Hesiod continues this tradition in *Theogony* and especially *Works and Days*; see further Taub (2003) 5–6, 10–11; Hannah (2005) 18–27.

[4] See e.g. Arat. 313–15 (Eagle); 744–7 (Arcturus). For regular reminders of the benevolence of Zeus and the profitability of the signs he sends, cf. esp. 5–14, 264–7, 740–57, 768–71. Moreover, on a few occasions, Aratus would seem to suggest that Zeus is also responsible for causing the weather in question (293, 886, 936). Zeus may represent here the providential creator of Stoic thought. But this may only be regarded as the predominant view in a poem which gives voice to competing systems of stellar thought; see Pendergraft (1990), Volk (2010) 200–4, and later discussion.

[5] See Volk (2010) 203–4, with later discussion. For ancient intellectual debates on the relationship between sign and signified, see Allen (2001). For traditional difficulties in determining significance and cause in weather folklore, see the interesting study by Ward (1968).

[6] Cf. Hom. *Il.* 22.30–1; Hes. *Op.* 587–8; Arat. 329–35; see elsewhere in the Greek tradition, Aes. *Ag.* 966–7; A.R. 2.516–17; see further McCartney (1926) 51–2.

[7] For the etymological connection between Sirius and σειριάει ('to burn'), cf. Arat. 329–32; cf. also Hyg. *Astr.* 2.35 *Sirion appellasse propter flammae candorem*; see further Le Boeuffle (1977) 134.

that all the stars exert individualised causal effects on atmospheric conditions.[8] Sections from Columella, Pliny the Elder or later writers are often cited in support of this view.[9] But these sources all date from the mid 1st century AD onwards and, though the idea of a causal relation between individual stars and weather patterns may be popular at this time, it would be rash simply to project this notion back into the minds of the thinkers of the previous centuries.

In the absence of any specific evidence to the contrary, I think one should take a literal approach to authors such as Hesiod and Aratus, and treat Sirius as a special case exception to a general view that the stars act merely as (divine) signs of weather, a unique circumstance that was undoubtedly influenced by the incredible brilliance of Sirius—it is the brightest star in the night sky—and its proximity to the sun.[10] One may find some additional support for the special place of the Dog Star in ancient Greek thought in an important section from the astronomical treatise of the 1st century BC Greek scientific writer, Geminos.[11] In Chapter 17 of his *Introduction to the Phenomena*, the author sets out to challenge the belief that the stars cause the weather (17.1):[12]

> Ὁ περὶ ἐπισημασιῶν λόγος παρὰ μὲν τοῖς ἰδιώταις ἀλλοίαν ἔχει διάληψιν ὡς ἐπὶ τῶν ἄστρων ἐπιτολαῖς καὶ δύσεσι τῶν περὶ τὸν ἀέρα μεταβολῶν γιγομένων. Ὁ δὲ μαθηματικὸς καὶ φυσικὸς ἑτέραν ἔχει δόξαν.

> The opinion concerning weather signs current among laymen involves a spurious belief that the changes in the air occur because of the [heliacal] risings and settings of the stars. But mathematical and physical opinion has it otherwise.

[8] See e.g. McCartney (1926) 53; Dicks (1963) 62–3; Hunter (1995); Taub (2003) 37; Hannah (2005) 19–20; Evans and Berggren (2006) 2; Gee (2007) 569; Lehoux (2007) 39.

[9] For a direct causal relationship between stars and weather, cf. e.g. Col. 11.1.32 *vim sideris*; Plin. *Nat.* 2.105–12, esp. 105 *quis enim aestates et hiemes quaeque in temporibus annua vice intelleguntur siderum motu fieri dubitet?*; 18.278–85, Gel. 14.1.9 (Favorinus).

[10] The outstanding brightness of Sirius was recognised by the ancients as early as Homer; cf. *Il.* 22.26–31. The exceptional status of Sirius among the constellations in Greek and Roman thought can also be evidenced by the fact that it warrants its own stellar harbinger, Procyon (lit. 'the one that goes before the Dog'); its special appearance on the Atlas Farnese; and the particular influence it was felt to have over fevers in the field of medicine; see further Brosch (2008) 20–7.

[11] For our information about Geminos and his works, and the general consensus that he was writing in the 1st century BC (at some point between 90 and 35 BC), see Evans and Berggren (2006) esp. 15–22.

[12] The text of Geminos is taken from Aujac (1975); translations are taken from Evans and Berggren (2006).

Having set out a general case against the idea (17.2–25), Geminos proceeds to give detailed attention to just one constellation, the Dog Star, in an exposition which takes up over half the chapter (17.26–45).[13] There is a tendency to treat this chapter in Geminos as evidence of a popular belief in the causal ability of *all* stars in the 1st century BC, with the Dog Star being used merely as a pertinent illustration.[14] Taken in the context of previous ancient views on the constellations, and with a view to what Geminos himself says elsewhere, I would suggest that it may be the Dog Star specifically which motivates the entire section. It is the tendency of all people (17.26), including poets and philosophers (17.32), to single out *this particular* constellation for special treatment, which motivates Geminos' detailed refutation in the chapter, as he rams home his central message that it is in fact the sun that is responsible for the hot weather wrongly attributed to Sirius (cf. 17.27, 30, 39, 45). The absence of other constellations from this chapter is telling, and reveals the very narrow focus of the issue that the author is attempting to address. Geminos' direct singling out of the Dog Star in this chapter is anticipated earlier in Chapter 3 when, in his brief listing of the zodiacal, northern and southern constellations, it is only in the case of Sirius that he draws attention (in sceptical fashion) to a popular belief in its direct influence over the weather (3.14):

> Ὁ δὲ ἐν τῷ στόματι τοῦ Κυνὸς λαμπρὸς ἀστήρ, **ὅς δοκεῖ τὴν ἐπίτασιν τῶν καυμάτων ποιεῖν**, ὁμωνύμως ὅλῳ τῷ ζῳδίῳ Κύων προσαγορεύεται.

> The bright star in the mouth of the Dog, **which seems to cause the intensification of the heat**, is called the Dog, in the same way as the whole constellation.

Once again, Geminos bears witness to the general tendency of writers and thinkers before him to treat the Dog Star as an exception to the rule, and the author feels a responsibility to persuade his readership to bring this constellation back into the fold with the others.

The current scholarly consensus, then, needs some corrective in my opinion. First, it seems to me that the extant record provides no convincing evidence for a belief in a strong causal connection between individual stars and specific weather patterns before the Roman writers

[13] This is not quite the case. There is mention of one other constellation in this section—the Pleiades (17.14)—but Geminos' comment is unclear: he appears (mistakenly) to cite Hesiod as evidence of a belief that the Pleiades caused rain.

[14] See Aujac (1975) 153; Kidd (1997) 411; Evans and Berggren (2006) 222.

of the 1st century BC.[15] I am not denying that a certain causal connection between stars and weather was entertained by Greek exponents of natural philosophy. Aristotle—across a number of his works, especially *Meteorology*—sets out a theory whereby hot weather on earth is attributed to the interaction between the *aither* and the (swift) motion of celestial bodies, especially in conjunction with the sun.[16] But Aristotle is concerned with the *motion* of the stars and the kinetic energy produced therein, rather than the particularisation of individual stars and their specific influence on all types of weather, not just heat—the topic that interests us here. To my mind, the change in attitude towards astrometeorology may be linked to the growing prominence of astrology in cultural and political circles at Rome from the late 1st century BC onwards. After all, if there is growing high-profile endorsement of the power of the constellations over the behavioural patterns and life expectancy of humans, it would hardly be surprising to find corresponding enthusiasm for a more directly causal role for them over the weather.

CICERO'S *ARATEA*: THE START OF A ROMAN ASTROLOGICAL TREND

We may begin with Cicero's translation of Aratus, which exists now in substantial fragments. In the light of Cicero's own testimony that he wrote the poem while he was still a young adult (*ND* 2.104 *admodum adulescentulus*), it is generally agreed that his translation of the first part of Aratus can be dated to *c.* 89–86 BC.[17] The dating of

[15] The figure of Arcturus at the beginning of Plautus' *Rudens* (1–82) does not compromise this position. It is true that Arcturus claims that he has directly caused the storm that motivates the action of the play; cf. *Rud.* 69 *increpui hibernum et fluctus movi maritumos*. But this is immediately contradicted within the play itself, where Neptune is held responsible for the storm (cf. *Rud.* 83–4, 358–60). Moreover, the same constellation also says that he, like other stars, has been appointed by Jupiter to act as moral adjudicator over mortals (*Rud.* 1–30). Arcturus is best treated as a comic braggart, a figure who starts by defining his status through civic association with Jupiter (*Rud.* 1–2) before deigning to claim for himself some of those same powers over land and sea; for further discussion of the theological/philosophical content of this prologue, see Fraenkel (1942); Hunter (1995).

[16] See esp. *On the Heavens*, 289a 19–35; *Meteorology*, 340b–341a; see further Taub (2003) 77–115, esp. 84–8.

[17] See e.g. Soubiran (1972) 8–9; Gee (2001) 520.

the entire project is complicated by a letter to Atticus of 60 BC (*Att.* 2.1.11), in which Cicero informs his friend of the imminent arrival of a work called *Prognostica*, which must refer to the second part of the Aratus translation. This may be the (long-awaited) continuation of a project which started almost thirty years previously or, perhaps more plausibly, a spare copy of the work sent to Atticus for him to consult alongside his reading of Cicero's recently composed and curiously Aratean work, *De Consulatu Suo*.[18] Either way, we can say with some confidence that Cicero's engagement with Aratus was a lot more than the isolated instance of youthful fancy that it is sometimes portrayed to be.

Close comparative analysis between Cicero and Aratus reveals the differing poetic priorities of the Roman translator, especially Cicero's desire to exploit the descriptive and anthropomorphic potential of the subject matter to create a more 'tactile' and dramatic cosmos.[19] But one conceptual difference between Cicero and Aratus has not yet been properly explored: namely, the stronger astrological import with which Cicero seems to endow his cosmos, as the major causal force in the creation of weather is transferred from the king of the gods (in Aratus) to the stars themselves.[20]

To start with, let us consider Cicero's description of Capricorn (*Arat.* fr. 33.57–9):[21]

> serius haec obitus terrai vissit Equi vis,
> **quam gelidum valido de pectore frigus anhelans**
> corpore semifero magno Capricornus in orbe;

> This powerful Horse visits the earth's curve (i.e. horizon) later than Capricorn, **which breathes out a frozen cold from its strong breast** when its semi-wild body is in the great circle (sc. of the Zodiac).

[18] For discussion of the possibilities, see Soubiran (1972) 9–16. For the *De Consulatu Suo* and its debt to Aratus, see Kubiak (1994).

[19] See esp. Toohey (1996) 79–87; Gee (2001) 523–7. This carries on a process inherited from Aratus himself, who enlivened in poetic fashion the material he took from the prose treatise of Eudoxus of Cnidus; see Kidd (1997) 14–18, 26–32.

[20] Lewis (1986) 220–1, 223–4, 227–8 draws attention to Cicero's downplaying of Jupiter in his version of Aratus, and attributes this to a preference for rationalising over theological views of the world that was prevalent among the Roman elite in the 1st century BC. But she does not consider this in conjunction with the more proactive role given to the stars in Cicero, which points us in a more astrologically focussed direction.

[21] The text of Cicero is taken from Soubiran (1972).

These lines represent a rendition of Aratus 283–6:[22]

πὰρ δ' ἄρα οἱ κεφαλῇ χεὶρ Ὑδροχόοιο
δεξιτερὴ τάνυται. ὁ δ' ὀπίστερος Αἰγοκερῆος
τέλλεται. αὐτὰρ ὅ γε πρότερος καὶ νειόθι μᾶλλον
κέκλιται Αἰγόκερως, ἵνα ἴς τρέπετ' ἠελίοιο.

Beside the Horse's head the right hand of the Water-pourer stretches out: he rises after Capricorn. Capricorn lies ahead and lower down, where the powerful sun turns back.

Whereas Aratus' interest here in Capricorn is purely in terms of relative spatial positioning, Cicero takes time to bring the sign to life as the monster that was translated to the skies. But this is also a constellation that is actively responsible for emitting icy exhalations, a sentiment which is most readily taken as an allusion to the wintry conditions that typically accompany its sign.

This is not an isolated instance of a greater role for the stars in the creation of weather. The Eagle, which is merely 'grievous/difficult' in Aratus, is specifically charged with churning up the sea in Cicero.[23] Among the list of constellations of the zodiac, the Crab, adorned with no epithet in Aratus, is given heat-bringing qualities in Cicero's translation.[24] In this conceptual context, it is not surprising to find that Cicero should elaborate on the destructive tendencies of the heat-bringing Dog Star, referred to in Latin by either *Canis* or *Canicula*: the six verses devoted to the subject in Aratus (329–35) are doubled in Cicero (*Arat.* fr. 33.107–19).

One further example is particularly important. Cicero speaks of the Altar as a constellation that mitigates the inclement weather (Cic. *Arat.* fr. 33.198–202):

sin gravis inciderit vehementi flamine ventus,
perfringet celsos defixo robore malos,
ut res nulla feras possit mulcere procellas,
ni parte ex Aquilonis opacam pellere nubem
coeperit et subitis auris diduxerit Ara.

But if a fierce wind shall have fallen upon [the sailor] with its savage blasts, it will shatter completely the high masts, firmly attached and

[22] The text and translation are taken from Kidd (1997).

[23] With Arat. 314 (χαλεπός), cf. Cic. *Arat.* fr. 33.89–90 *[Aquila] . . . sed grave maestis/ ostendit nautis perturbans aequora signum* ('but Aquila shows a severe sign to wretched sailors as it churns up the waters').

[24] With Arat. 545, cf. Cic. *Arat.* fr. 33.320 *aestifer . . . Cancer.*

strong though they be, such that nothing will be able to soothe its fierce storms, unless, from the region of the North (wind), the Altar shall have begun to dispel the dank cloud and shall have dispersed it with sudden breezes.

This is in fact a very free expansion of Aratus 423–39, which may reveal our author's greater tendency to think in astrological terms. In Cicero, it is the constellation which comes to the sailor's rescue in a tempest, whereas the role of saviour in Aratus is expressly reserved for the benevolent Zeus (Arat. 426).

Cicero's decision to deny the king of the gods his Aratean omnipotence over the weather may have been a consistent feature of his poem. Aratus' description of the seven Pleiades (254–67) ends with a positive endorsement of Zeus' overriding influence (264–7):

αἱ μὲν ὁμῶς ὀλίγαι καὶ ἀφεγγέες, ἀλλ' ὀνομασταὶ
ἦρι καὶ ἑσπέριαι, **Ζεὺς δ' αἴτιος**, εἱλίσσονται,
ὅ σφισι καὶ θέρεος καὶ χείματος ἀρχομένοιο
σημαίνειν ἐπένευσεν ἐπερχομένου τ' ἀρότοιο.

All alike [the Pleiades] are small and faint, but they are famous in their movements at morning and evening, **and Zeus is the cause**, in that he authorised them to mark the beginnings of summer and winter and the onset of ploughing time.

Cicero's account of the same constellation gives centre stage to the stars themselves (*Arat.* fr. 33.37–41):

hae tenues parvo labentes lumine lucent;
at magnum nomen signi clarumque vocatur,
propterea quod et aestatis primordia clarat,
et post, hiberni praepandens temporis ortus,
admonet ut mandent mortales semina terris.

Weak are these stars as they shine, gliding along, with little light; but great and distinguished is the name by which the constellation is called, on account of the fact that it declares the beginning of summer and, after that, revealing in advance the rise of winter, it warns mortals to commend their seeds to the lands.

The contrast between their reputation and the light they emit, and their role as indicators of summer, winter and the time for ploughing are all retained from Aratus—but Ζεὺς δ' αἴτιος has gone, as the Ciceronian stars become the subject of active verbs. In fact, in all five references to Jupiter in the extant Ciceronian poem (*Arat.* fr. 1, fr. 19, fr. 33.20, 188, 294), none of them pay any lip service to his Aratean role as benevolent sender of signs to mortals.[25]

Now, it is important not to overstate the strength of Cicero's astrological reimagining of Aratus. The examples given are, after all, subtle and sporadic. Nor does the very poorly preserved state of Cicero's *Prognostica*—the most natural place to construct a case for stellar causation with regards to weather—allow us to venture further about the possible astrological ambitions of our author: the remaining fragments certainly make no mention of any constellation.[26] We also need to consider the verdict on the poem made by Cicero himself (via Quintus) when he seems to imply that his Aratus dealt with *signs* rather than *causes*.[27] Moreover, even if we accept a stronger astrological thrust to Cicero's poem, we might argue for a slightly softer transition between source and Latin rendition and suggest that Cicero is giving considerable expansion to a notion which is latent in Aratus.[28]

[25] Nothing remains of Cicero's version of the Aratean prologue, which would have been the most obvious place to signal a change in emphasis away from an omnipotent Zeus. Particularly telling is *Arat.* fr. 33.155–66, where Cicero acknowledges the mortal role in the naming and shaping of the stars (158–9, also 162 *astrorum custos*) but substitutes the Aratean Zeus for Nature (160 *natura*) as the supernatural influence.

[26] A curious fragment of astronomical poetry, preserved in Ausonius, is attributed in the superscription to Cicero's brother Quintus; see Courtney (1993) 179–81. Lines 1–13 deal with the weather connected with the signs of the zodiac, and there are noticeably strong causal links created between them: cf. 1 *flamina verna cient obscuro lumine Pisces* ('The Fishes with their dim light summon the spring breezes'); 6 *languificosque Leo proflat ferus ore calores* ('The fierce Lion breathes forth sloth-making heat from its mouth'); 7 *post modium quatiens Virgo fugat orta vaporem* ('Then the Maiden rises and, shaking her measuring-vessel, puts the heat to flight'); 10 *ecfetos ramos denudat flamma Nepai* ('The flame of the Scorpion strips bare the worn out branches'); 11 *pigra Sagittipotens iaculatur frigora terris* ('The Archer hurls sluggish cold on to the lands'); 13 *quem sequitur nebulas rorans liquor altus Aquari* ('The deep water of Aquarius follows him, bedewing the clouds'). It is impossible to know whether this is indeed a (more astrologically oriented) version of Aratus from Cicero's brother, possibly dating from 45–43 BC (see Possanza 1992) or an additional fragment, or rather two additional fragments, to be added to Cicero's poem (see Gee 2007). Lines 1–13 show a very well-developed astrological system, way beyond anything seen in Cicero and closer in overall spirit to Germanicus (see later). For this reason, I would be more ready than Gee (2007: 578 n. 60) to entertain attributing it to Germanicus.

[27] Cf. *Div.* 1.13–15, esp. 13 *atque his rerum praesensionibus prognostica tua referta sunt. quis igitur elicere causas praesensionum potest?* ('your *Prognostica* was stuffed full of warnings of these things: who therefore can work out the causes of these warnings?'). At any rate, the decline in enthusiasm for stellar causation from his *Aratea* to *De Divinatione* is interesting, and lends additional weight to the view that the comments in *Div.* are motivated by the immediate political circumstances; see Chapter 4.

[28] There is one hint of a constellation other than the Dog Star causing weather in Aratus, namely the Kids, which at Arat. 682, are felt to 'set in motion' storms (κινῆσαι); see Kidd (1997) ad loc.; Martin (1998) ad loc.; Volk (2010) 204.

Nevertheless, the greater range of responsibilities given to the stars themselves in Cicero's poem, combined with a conscious downplaying of the role of Jupiter, demand attention. What we may be witnessing here is a starting point, a willingness on the part of intellectual Romans to give greater respect to a more astrological interpretation of the relationship between stars and weather. It is worth recalling that the timing of Cicero's poem in the early-to-mid 80's BC coincides with some of our earliest anecdotal evidence for high-profile Roman engagement with astrology. What better way could there be of signalling this new emphasis than with an 'astrological updating' of a text that was very widely read by the Romans?[29]

CAESAR'S *DE ASTRIS*: THE HARDENING OF THE ASTROLOGICAL LINE?

As we move towards the imperial period, it is astrometeorology, one could argue, which first achieves prominent political endorsement and exposure, in the form of *De Astris*, a treatise, now lost to us, which appears to have been at least partly composed by Julius Caesar in connection with the reformation of the Roman calendar.[30] Whatever its precise nature, this treatise was evidently a major influence on future astrometeorological writers: Julius Caesar is unambiguously referred to as a source of stellar information by Pliny the Elder; Ptolemy, Johannes Lydus and the writers of the al-Biruni *parapegma* may also have been influenced, if one accepts the most likely view that the generic *Caesar* of these sources refers to Julius, rather than Germanicus and his *Phaenomena*.[31]

[29] For the view that Aratus was a standard school text, see Lewis (1992) 113–17.

[30] There is general scholarly consensus around the view that *De Astris* was a joint venture between Julius Caesar and his Egyptian astronomer, Sosigenes, who was responsible for the astronomical calculations; see e.g. Le Bonniec and Le Boeuffle (1972) 28–9; Fantham (2009) 153–5.

[31] In *Nat.* 18, Pliny outlines Julius Caesar's reformation of the calendar (211) before revealing his intention to pay particular attention to Caesar's reckonings for the dates for the rising and setting of the stars (214 *nos sequimur observationem Caesaris maxime; haec erit Italiae ratio*); references to Caesar duly follow in sections 234, 237, 246–8, 255–6, 268–71 and 309–11 of this book; Ptolemy's *Phaseis* (cf. e.g. Phaophi 26 (Lehoux 2007: 264), Athyr 4 (Lehoux 2007: 287), Choiak 5 (Lehoux 2007: 290), Phamenoth 27 (Lehoux 2007: 298)); Lydus, *De Mensibus* (cf. 4.18, 4.128); for the al-Biruni *parapegma*, see Lehoux (2007) 430–70. It is not only Julius Caesar's work on the calendar that makes him the more likely candidate for the *Caesar* referred to in these sources: Germanicus' work is for the most part a version of Aratus, an author who is nowhere cited as an authority in the extant *parapegmata* (see Lehoux 2007: 492–4).

In the light of Pliny the Elder's admission that he is mostly following the observations of Julius Caesar (*Nat.* 18.214), it is tempting to use Pliny's text to try to ascertain the nature of the lost work.[32] If we pursue this strategy for a while, we could make the following findings. In some respects, it would appear that Caesar's approach was traditional. Several examples could be cited where Caesar appears to adhere to the traditional view of the constellations as signals of forthcoming weather. Take, for example, *Nat.* 18.237:[33]

> a favonio in aequinoctium vernum Caesari significant, XIII kal. Mart. triduum varie, et VIII kal. hirundinis visu et postero die Arcturi exortu vespertino, item III. non Mart.
>
> Between the period of the west wind and the spring equinox, 16 February for Caesar marks three days of changeable weather, as also does 22 February by the appearance of the swallow and on the next day by the rising of Arcturus in the evening, and the same on 5 March.

Moreover, the Dog Star continues to be treated as a special case (*Nat.* 18.269–72):

> XVI kal. Aug. Assyriae Procyon exoritur, dein postridie fere ubique confessum inter omnes sidus indicans, quod Canis ortum vocamus . . . sentiunt id maria, terrae, multae vero et ferae, ut suis locis diximus. neque est minor ei veneratio quam descriptis in deos stellis, accenditque solem et magnam aestus obtinet causam . . . in hoc temporis intervallo res summa vitium agitur decretorio uvis sidere illo, quod Caniculam appelavimus . . .
>
> On 14 July Procyon rises for Assyria, and then the day after, almost everywhere, reveals itself, a constellation which is universally acknowledged, which we call the rising of the Dog Star . . . The seas and the lands feel it, and indeed many wild animals besides (as we have said in the proper places). Nor is less reverence owed to it than to the stars that are allotted to the gods [i.e. the planets], and it kindles the sun and constitutes a great source of heat . . . During this interval occurs the critical time for the vines, as that constellation which we have called the Little Dog stands as judge over the fate of the grape.

The Dog Star's causal influence over hot weather is emphatically expressed in terms of its actively kindling the sun, and the special

[32] The following is influenced by the interesting analysis of Cramer (1954: 75–7), though I think Cramer is overly optimistic in his attempts to recreate the spirit of the lost work.

[33] The text of Pliny is taken from Le Bonniec and Le Boeuffle (1972).

reverence it receives above all else, even the stars of the gods, for its power in dispensing judgement over nature.[34]

But there is also a stronger, astrological thrust to the astrometeorology we find in Pliny. Take, for example, the discussion of the Pleiads (*Nat.* 18.225):

> servantque id sidus etiam vestis institores, et est in caelo notatu facillimum. ergo ex occasu eius de hieme augurantur quibus est cura insidiandi, negotiatori avaritia. nubilo occasu pluviosam hiemem denuntiat, statimque augent lacernarum pretia; sereno asperam, et reliquarum vestium accendunt.

> Even clothes dealers observe that constellation and it is easiest to identify in the sky. Consequently, those whose concern it is to seize upon the favourable moment, spurred on by the greed of a tradesman, make forecasts about the winter from its setting. By a cloudy setting it foretells a wet winter, and they at once increase the prices for cloaks; whereas by a fine setting it foretells a harsh winter, and they inflate the prices of all other clothes.

Traditionally, astrometeorology had always implicitly spoken to the lives and livelihoods of specific mortals: farmers and sailors. Pliny here extends this tradition to include more general retailers, and he is much more explicit about the financial benefit that attends a successful reading of the stars. In extending the role of the stars over the behavioural patterns of humans more generally, Pliny's astrometeorology blends with the catarchic form of astrology peddled by astrologers in the streets of Rome.

In all these examples from Pliny, of course, it is impossible to tell whether the sentiments derive from the author or his source: does the stronger astrological thrust to the astrometeorology come straight from Julius Caesar, or is this Pliny embellishing his narrative at a time when horoscopic astrology has become well established in the Roman Empire? We cannot be sure, but it is worth bearing in mind the possibility that Caesar is, like Cicero before him, extending the remit of astrometeorology, especially when we come to deal with the (undeniably) strong astrological charge that is given to the subject by his great grandson of the imperial line, the prince Germanicus.

[34] *decretorius*, a rare adjective first occurring in Seneca, is found outside Pliny to refer to crucial decision-making in the philosophical or military spheres; cf. Sen. *Clem.* 1.14.1; *Ep.* 102.24, 117.25; Quint. 6.4.6, 10.5.20; *TLL* 5.1.220.59–67. Pliny himself uses it elsewhere only at *Nat.* 17.11 and 18.288.

STARS FOR IMPERIAL WEATHER: BUILDING UP TO GERMANICUS' *PHAENOMENA*

In the light of the influence of Aratus and Hellenistic poets in general, and contemporary interest in stars in Augustan Rome, it is not surprising to find Augustan writers engaging with astrometeorology in various contexts.

Several Roman poets adopt what we might call the traditional approach to the role of stars in weather prediction, including the special powers attributed to the Dog Star. Virgil in his *Georgics*, written in the 30's BC, looks back to both Hesiod and Aratus in regarding constellations as signs of seasonal progression which implicitly inform farmers what particular chore they need to perform.[35] Consistent with Aratus, it is Jupiter who is deemed controller of both the weather and the guiding signs.[36] Hesiodic influence is particularly felt in Virgil's emphasis on the importance of stars for both farmers and sailors.[37] Moreover, Virgil singles out the Dog Star for special treatment as an active force in the creation of hot weather.[38] Both Propertius and Tibullus have less to say about constellations, and yet both single out Sirius for special treatment as an active force in weather-making, practically equating it with the sun.[39] Ovid is, if anything, more circumspect: phraseology

[35] Cf. Verg. *G.* 1.1–2 *quo sidere terram vertere*, 67–8, 73, 208–30, 4.231–5; for the general principle of constellations as guidance to weather/chores, cf. 1.252–8 with Arat. 7–9; see further Getty (1948); Taub (2003) 54–8.

[36] Cf. Verg. *G.* 1.351–5.

[37] Cf. Verg. *G.* 1.204–14, 253–5.

[38] Cf. Verg. *G.* 2.353 *ubi hiulca siti findit Canis aestifer arva* ('when the heat-bringing Dog splits the fields which gape with thirst'). The use of *aestifer* here directly recalls Cicero, who may have coined the adjective and is the first to apply it to stars at *Arat.* fr. 33.320 Soubiran; see *TLL* 1.1091.74–81. But it is noticeable that Virgil transfers the heat-bringing quality of stars from Cancer, as in Cicero, back to the more traditional Dog Star. Cf. also 4.425–6 *iam rapidus torrens sitientis Sirius Indos/ardebat caelo* ('and now Sirius, swiftly parching the thirsty Indians, was burning in the sky'); also *A.* 3.141 *tum sterilis exurere Sirius agros* ('then Sirius burned the barren lands'), 10.273–4 *Sirius ardor/ille sitim morbosque ferens mortalibus aegris* ('that burning of Sirius which brings thirst and diseases to ailing mortals').

[39] Cf. Prop. 2.28A.4 *incipit et sicco fervere terra Cane* ('the earth begins to heat up with the dry Dog Star'); Tib. 1.1.27–8 *sed Canis aestivos ortus vitare sub umbra/arboris* ('but avoid the summer rising of the Dog Star beneath the shade of a tree'), 1.4.6, 42 *et Canis arenti torreat arva siti* ('and the Dog Star scorches the fields with burning thirst'), 1.7.21 *arentes cum findit Sirius agros* ('when Sirius cleaves the burning fields'); cf. also Lygd. 3.5.2.

such as *(ad)moto sidere* in references to Sirius indicates temporal coincidence with, rather than active influence over, hot weather.[40]

Against this backdrop, Horace emerges as the most obvious ambassador for the new type of astrometeorological thinking, as he combines approaches old and new. Horace regularly attests to the traditional notion of constellations as indicators of weather.[41] He also follows tradition by attributing to the Dog Star a strong causal effect on the hot weather on earth.[42] But Horace also, like Cicero, extends the causal role of stars to cover constellations other than the Dog Star. Arcturus is felt to be actively attacking the mortal world with its storms.[43] More clearly, and generally reminiscent of Cicero's translation, Orion is made the subject of an active verb as it is deemed directly responsible for churning up the sea.[44] And a grouping of apparently active stars is seen to collude to cause a drought (*Carm.* 3.29.17–20):

> iam clarus occultum Andromedae Pater
> ostendit ignem, iam Procyon furit
> et stella vesani Leonis,
> sole dies referente siccos.

> And now the distinguished Father of Andromeda (Cepheus) shows his hidden fire, now Procyon rages and the star of maddened Leo, as the sun brings back the days of drought.

Though Horace strictly attributes drought here to the sun only, the references to Cepheus' fire and the fury of both Procyon and Leo suggest that he envisages a causal connection between all four celestial bodies

[40] Cf. *Am.* 2.16.3–4, *Fast.* 4.939–40.

[41] For example, the south wind merely accompanies Orion (*Carm.* 1.28.21–2 *rabidus comes Orionis . . . Notus*), and only a temporal connection is acknowledged between weather and the constellations of the Pleiades (*Carm.* 4.14.21).

[42] Cf. *Carm.* 1.17.17–18 *hic in reducta valle Caniculae/vitabis aestus* ('here in a secluded valley will you avoid the heat of the Little Dog'). In other references, Horace does not specifically refer to the Dog Star, though the special status of the constellation has become so conventional by now that the author can evoke it simply by references to heat: *torrentia . . . sidera* (*Carm.* 3.1.31–2); *siderum . . . vapor siticulosae Apuliae* (*Epod.* 3.15–16); *astri . . . aestuosa . . . impotentia* (*Epod.* 16.61–2); cf. also *sidus fervidum* (*Epod.* 1.27).

[43] Cf. *Carm.* 3.1.27–8 *saevus Arcturi . . . impetus*. Other constellations in Horace are attributed adjectives which suggest their active force on the weather; cf. *furibundus*, used of Leo at *Ep.* 1.10.16–17.

[44] With Cic. *Arat.* fr. 33.89–90 Soubiran (Aquila), cf. *Epod.* 15.7–8 *dum . . . nautis infestus Orion/turbaret hibernum mare* ('so long as . . . Orion, hostile to sailors, was tossing the winter's sea').

and the unpleasant weather. As was the case in Cicero, the Dog Star now has other active stellar colleagues with which to contend.

To attribute the Horatian variety simply to poetic licence would be misplaced: if this is the case, are we implicitly judging Hesiod, Aratus and indeed Virgil as lesser poets for not similarly varying their own long accounts in this way? Rather, the primary motivation behind Horace's decision to extend the causal role of stars beyond the Dog Star lies in the contemporary Roman cultural context and the growing power of the stars under the empire.[45]

If the winds of change are felt in Cicero and Horace, they blow more forcefully with Germanicus. Germanicus, grandson of Augustus by adoption from AD 4, has left us a poetic rendition of Aratus.[46] The extant portion of Germanicus' poem spans 725 lines, which adhere closely to the first part of Aratus' *Phaenomena* (1–731).[47] But there are some notable divergences which help Germanicus to strike a suitably Roman imperial note. The opening invocation, in fact, immediately signals a departure from his Hellenistic source (1–16):[48]

> ab Iove principium magno deduxit Aratus.
> carminis at nobis, genitor, tu maximus auctor,
> te veneror tibi sacra fero doctique laboris
> primitias. probat ipse deum rectorque satorque.
> quantum etenim possent anni certissima signa,
> qua Sol ardentem Cancrum rapidissimus ambit
> diversasque secat metas gelidi Capricorni
> quave Aries et Libra aequant divortia lucis,

[45] I agree, with some reservation, with the analysis of Dicks (1963: 62–5).

[46] I am here following the widely held scholarly viewpoint that the poem we possess can be attributed to Germanicus; for discussion, see esp. Possanza (2004) 219–33. The date parameters for Germanicus' poem are AD 4–19, from the time of his adoption into the imperial household along with Tiberius (to whose dynastic connections the author appears to allude in 1–16) to his death. Beyond this, there continues to be scholarly debate, depending on whether one believes the work to be written under Augustus or Tiberius, and whether or not the reference to Augustus' apotheosis (558–60) is viewed as a later addition to an essentially Augustan text. For a late, Tiberian date (*c.* AD 14–19), see e.g. Le Boeuffle (1975) vii–x; Santini (1977) 32; Cicu (1979); Zehnacker (1989) 318–24; for an (essentially) Augustan dating for the poem, see e.g. Fantham (1985) 254–6; Possanza (2004) 233–5; Volk (2009) 189 n. 29. For my purposes, once again, it is the imperial nature of the addressee, rather than his specific identity, which is important.

[47] For comparative overviews, see Le Boeuffle (1975) 75–6; Possanza (2004) 8–9. We are not in a position to assess the interaction between Germanicus and Ovid's own *Phaenomena*, of which only two fragments survive; see Courtney (1993) 308–9.

[48] The text of Germanicus is taken from Gain (1976).

si non parta quies te praeside puppibus aequor
cultorique daret terras, procul arma silerent?
nunc vacat audacis ad caelum tollere vultus
sideraque et mundi varios cognoscere motus,
navita quid caveat, quid scitus vitet arator,
quando ratem ventis aut credat semina terris.
haec ego dum Latiis conor praedicere Musis,
pax tua tuque adsis nato numenque secundes.

From mighty Jupiter did Aratus launch his beginning. But for the beginning of my poem you, father, are the greatest instigator: it is you to whom I pay honour, to you I bring sacred gifts, the first-fruits of my learned labour. The ruler and begetter of the gods himself approves. For what power would there be in the most regular constellations of the year, where the sun travels most rapidly around the burning Cancer and grazes the opposite turning-post of icy Capricorn, or where the Ram and the Balance make equal the divisions of the day, if the creation of peace under your guardianship were not giving the sea to the ships nor the lands to the farmer, if arms were not now silent and distant? Now there is the opportunity to raise one's brave face to the sky and recognise the stars and the varied movements of the universe, of what the sailor should take heed and what the experienced ploughman should avoid, when the sailor should entrust his bark to the winds or the ploughman his seeds to the lands. While I try to foretell these things with Latin Muses, may you, your peace and your divinity be present and favour your son.

In marked contrast to Aratus, whose didactic treatise on the constellations is not tied to any particular time, Germanicus' poem is set firmly in early imperial Rome and celebrates the opportunity for farmers and sailors to renew their stellar observations in the light of the specific peace brought about on land and sea by Augustus (5–14). In the process, Germanicus reconfigures the ideological thrust of Aratus' poem: in place of a theistic vision of an all-powerful and benevolent Zeus, emphasis is placed on the peace-bringing benevolence of Augustus, the earthly counterpart to Jupiter, in allowing renewed contemplation of the stellar world.[49]

Augustus is later evoked with regard to a (complex) celebration of his apotheosis (554–60):

[49] See further Lewis (1986) 212–15; Possanza (2004) 105–16. The fact that we no longer possess Cicero's version of the Aratean prologue means that we cannot assess the extent to which Germanicus may have been influenced by the earlier Latin translator.

cochlidis inventor, cuius Titania flatu
proelia commisit divorum laetior aetas,
bellantem comitata Iovem, pietatis honorem
ut fuerat, geminus forma, sic sidere, cepit.
hic, Auguste, tuum genitali corpore numen
attonitas inter gentis patriamque paventem
in caelum tulit et maternis reddidit astris.

The inventor of the conch—by whose blast a happier age of gods accompanied warring Jupiter and engaged in battle with the Giants—received honour by virtue of its devotion: and as it had been on earth, twin-formed, so it is as a star. This sign, Augustus, carried your godhead away from your engendered body[50] and into the sky amid the quaking nations and frightened fatherland, and returned it to the mother star(s).

The *cochlidis inventor* (554) is Capricorn, the advertised star sign of Augustus, and specific recollection of its role in the gigantomachy creates a flattering connection between Emperor and constellation: it is intimated that Augustus, just like Capricorn, is being rewarded in heaven for his pious involvement in bringing to an end a war against evil (i.e. Actium). What the author means in 558–60, however, is subject to dispute. Is he talking about a catasterism, a transformation into a star, or is this simply a case of Augustus' divine spirit (*numen*, 558) returning to the heavenly place from where it originated? And who or what is/are the *materna astra* (560): Capricorn? Libra? a generic reference to the birthplace of Augustus' heavenly spirit?[51] Whatever the precise implication of 558–60—be it deification or catasterism—these lines emit a strong, imperially sanctioned astrological charge.

Imperial interests are also served in parts of the work which contain no direct reference to the Emperor.[52] For our purposes, the most

[50] I follow Gradel (2002: 284–5) in interpreting *genitali corpore* as a separative ablative.

[51] For differing views, see Montanari Caldini (1976) 48; Santini (1977) 19–22; Barton (1995) 36; Domenicucci (1996) 109–10; Possanza (2004) 182–4.

[52] See esp. Lewis (1986) 221–2, 224–5, 228–30. Moreover, Le Boeuffle (1983b: 91) astutely draws attention to the positive dynastic ideology contained within the stories of Cepheus and Perseus: see esp. 186 *prodest maiestas saepe parentis* ('the majesty of one's parent is often an advantage'), 250 *moles ipsa viri satis est testata parentem* ('the very weight of the man is testimony enough of the parent'). It must be acknowledged, however, that Germanicus is not wholly successful in his attempt to produce a Roman imperial vision of Aratus. In particular, it is unfortunate that his version of the myth of the Maiden (96–139), which adheres to the degenerative account of ages found in Aratus, makes no concession to the *pax Augusta* celebrated in the introduction, or indeed, to the 'renewed' Golden Age of imperial discourse.

interesting alteration which Germanicus makes to his Aratean model occurs when he replaces Aratus' refusal to deal with the planets (Arat. 454–61) with an oblique promise to handle them elsewhere if the time and effort permit (437–45, esp. 444–5):[53]

> hoc opus arcanis an credam postmodo Musis,
> tempus et ipse labor, patiantur fata, docebit.

> Whether I can a little later entrust this work to the Muses that hide away,
> time and effort itself will show, fates permitting.

These lines would appear to look forward to a second part to the poem, which now only exists in five fragments. Fragments 2, 3 and 4 pertain to the influence of the stars on earthly activities. Fragment 2 (*c.* 20 lines) deals with the speed of movement of the planets. Fragments 3 and 4 are more substantial—together amounting to just over 190 lines—as they deal first with the general influence of the zodiacal constellations over the weather (fr. 3.1–22) before moving on to the effects that each planet has when it enters a particular sign (fr. 3.23–4.163).

The detail of these fragments does not correspond to anything specifically in Aratus—nor indeed anything in previous extant literature—but it might be taken as a strongly astrological reinterpretation of the second part of Aratus' poem (*Prognostica*). Germanicus has already replaced the all-powerful Aratean Zeus with Augustus in his version of the poem, and this has consequences for how we are to view the stars. No longer the control of an overarching deity like Zeus, the constellations and (deified) planets—listed as just one player amongst which is Jupiter (fr. 3.4, fr. 4.1–24)—are all conscious entities in themselves, specifically responsible for causing atmospheric conditions on earth.[54]

One section may be taken as representative of Germanicus' overall style (fr. 3.10–22):

> Virgo **refert** pluvias et **permovet** aera ventis.
> lenius est Librae signum; vix rorat in illo.
> Scorpios assidue caelo **minitabitur** ignis
> atque truces ventos; densa nive saepe rigebunt.
> rara Sagittifero descendunt flamina terris;
> lentior in pluvias, magis invenit arva quieta.

[53] An intention to deal with planets may be signalled as early as the invocation, when the author relishes the opportunity to learn the 'varied movements of the universe' (12 *mundi varios cognoscere motus*).

[54] With discussion later, see also Montanari Caldini (1973), (1976); Le Boeuffle (1983b); Abry (1993) 187–9.

Aegoceros alias **parcit**, sed frigora **durat**,
instabilita gelu falli vestigia passus.
qui fundit latices caelo quoque **permovet** imbris.
omnia miscentur cum Piscibus; aspera ventis
aequora turbatos volvunt ad sidera fluctus;
imbribus incumbit caelum solemque recondit;
grandine pulsatur tellus, nive condita dura est.

The Maiden **brings back** the rain waters and **throws into commotion** the air with the winds. The Balance is a gentler sign; scarcely does dew fall under it. The Scorpion continuously **threatens** the sky with fires and fierce winds; they are often stiff from the dense snow. Under the Bow-bearer breezes seldom descend to earth; it is milder with respect to rain, and finds the fields more quiet. In certain respects Capricorn **is sparing**, but **it hardens** the cold, permitting men's feet to become unsteady and slip on the ice. The being who pours forth water in heaven also **throws into commotion** the rains. With the Fishes all things are mixed together: the harsh sea rolls to the stars its waves which have been churned up by the winds; the sky labours under showers and hides the sun; the earth is struck by hail and is buried in hard snow.

These lines are typical of Germanicus' style and strategies in fragments 3–4. First, Germanicus lends the discussion of astrometeorology a distinctly astrological feel by dealing principally with the zodiacal constellations and the planets, the heavenly entities most important to astrology. Secondly, the author gives a much greater causal sense to the stars' involvement in the formation of weather by explicitly making most of the constellations subjects of active verbs. The traditional causal power of the Dog Star now has to jostle for influence among his competitors, such as the planetary Mars (fr. 4.40–1):

invectus Cancro, terras cum letifer ortu
Sirius afflavit, nocituros temperat aestus.

Carried into the Crab, at the time when death-bringing Sirius has breathed onto the lands at its rising, [Mars] tempers the heat designed to cause harm.

Moreover, in this section, Germanicus seeks to justify the causal connection by alluding to the mythical interpretation of the constellations: so, for example, the Scorpion is seen in threatening pose, just as it is on earth, and Aquarius, identified here as the heavenly water-pourer Ganymede, inflicts the same activity on mortals.

Scholars duly note that Germanicus provides us with our first extant instance of an astrological, planetary approach to the relationship between the stars in heaven and atmospheric conditions on earth. A similar astrological approach becomes popular with later Roman commentators, such as Columella and Pliny the Elder, and it is found in later Greek astrological writers, especially Ptolemy in the *Tetrabiblos* and, more sporadically, in Vettius Valens.[55]

But there is no need to assume, as is the tendency of scholars in this area, that Germanicus is tapping into an existing idea and working from a source now lost to us. There are sufficient examples from the author's eminent Latin predecessors—Cicero, Horace and perhaps his own grandfather Julius Caesar—to argue for a gradual (if not necessarily linear) progression of astrological thought from the early 1st century BC onwards, a progression which finds its counterpart in the intellectual and political sphere at Rome. To my mind, Germanicus' primary motivation here is political, a wish to celebrate the powerful place of astrology in the life of his poem's imperial addressee.

But it is equally important not to overlook the significance of the prince's discretion in astrological matters. Germanicus avoids totally any discussion of the branches of astrology which impinge more directly on the physical, moral or behavioural patterns of individuals, such as catarchic or natal astrology.[56] With hindsight, one can appreciate more fully the importance of Germanicus' apparently unparalleled epithet for the Muses. *arcanis . . . Musis* (444), 'Muses that hide away/ keep silent', is a striking collocation which speaks against the usually close relationship between the figure of the poet and his Muses.[57] Contrary to what was said explicitly in 444–5, the ability to deal with all aspects of astrology is evidently not simply a matter of the poet's time.

To summarise, my central point in this chapter is to argue that, contrary to common scholarly opinion, one can detect in the extant record a development in the role of stars in Roman astrometeorological thinking, from signs of forthcoming weather—with the Dog Star as a notable exception—to causal influences on the atmosphere. And

[55] For Roman writers, see n. 9 earlier in this chapter. See also Abry (1993) 187–9.

[56] On my reading, it would be misplaced to suggest that a treatment of horoscopic astrology may have been included in the parts which are now lost (an idea floated by Le Boeuffle 1983b: 91).

[57] *arcanus* in an active sense is rare, and the *TLL* offers no parallel for its use as an epithet for the Muses; see *TLL* 2.434.55–75.

the thrust of this change occurs during a period at Rome in which astrology is becoming embedded in intellectual and political discourse more generally.

Moreover, in terms of acceptable parameters of stellar investigation in the imperial age, Germanicus can be seen to be in dialogue with Vitruvius. Like Vitruvius, imperial catasterism is endorsed, on this occasion in the case of Augustus. And experimentation with Hellenistic astronomical texts continues, as Germanicus engages consciously with both Aratus and Eratosthenes: the frame of Aratus' *Phaenomena* is supplemented with many additional aetiological enquiries into catasterisms drawn from Eratosthenes.[58] But Germanicus refines the implicit directives laid down by Vitruvius by judging astrometeorology a worthy and politically harmless means of celebrating the power of the stars, even if natal astrology remains strictly off limits.

Another worthy instance of stellar celebration was found in the form of Caesar's catasterism, a topic to which we turn in Chapter 9.

[58] Whereas Aratus has a predominantly 'astronomical' outlook in his *Phaenomena*—only 14 of the 46 constellations he mentions attract mythical comment—Germanicus significantly elaborates on the mythical background to the stars; see esp. Possanza (2004) 170–217.

9

Caesar's Comet: The Reinvigoration of a Religious Enquiry

This chapter is dedicated to contemporary Augustan literary treatments of the comet of Caesar that appeared in all its glory in July 44 BC.[1] A good starting point for the discussion is provided by Ovid, who offers the most elaborate rendition of the apotheosis of Caesar in the final 'historical' episode of his *Metamorphoses*, as one of a series of deification stories in the latter part of the epic.[2] The poet picks up the story close to the time of the murder of Caesar and narrates the heavenly attempts of Venus to prevent the imminent assassination of her beloved progeny. After a seemingly unsuccessful petition to the gods (15.765–78), Venus tries to protect the victim from the onslaught (15.799–806), before Jupiter reveals to her that Caesar is fated both to lose his life now and find a place in heaven as a god (15.807–42). As instructed, Venus descends to earth to pick up the soul of the deceased leader and carry it off to heaven, during the course of which Caesar's soul catches fire and creates, from the perspective of the mortal onlooker, the appearance of a comet (15.843–51).[3]

[1] An important qualification must be made immediately. I refer to Caesar's comet here because I follow the prevailing view that the comet was a historical event, despite the silence of the contemporary evidence (see Chapter 3). But I am not suggesting that all Augustan commentators handle this astronomical phenomenon as if it were a comet: as will become clear, I concur with those scholars who see a development in the identification of this stellar phenomenon.

[2] Cf. the apotheoses of Hercules (9.262–72), Aeneas (14.581–608), Romulus and Hersilia (14.805–51); Ovid also looks forward to the apotheoses of Augustus (15.852–70) and indeed himself (15.871–9).

[3] In so doing, the epic myth has conflated the historical time between the death of Caesar and the sighting of the comet during his funeral games some months later.

While Ovid's story is by no means deficient in the sort of grand epic motifs that befit the political subject matter, scholarly attention has long been focussed around the ways in which the myth invites critical reflection on the political machinations behind Caesar's deification, creating tensions between reality and imperial rhetoric.[4] The comparison between Julius Caesar and his illustrious son at the beginning of the story exposes such tensions immediately (15.746–61):[5]

> Caesar in urbe sua deus est; quem Marte togaque
> praecipuum non bella magis finita triumphis
> resque domi gestae properataque gloria rerum
> in sidus vertere novum stellamque comantem,
> quam sua progenies. neque enim de Caesaris actis
> ullum maius opus quam quod pater exstitit huius.
> scilicet aequoreos plus est domuisse Britannos
> perque papyriferi septemflua flumina Nili
> victrices egisse rates Numidasque rebelles
> Cinyphiumque Iubam Mithridateisque tumentem
> nominibus Pontum populo adiecisse Quirini
> et multos meruisse, aliquos egisse triumphos,
> quam tantum genuisse virum? quo praeside rerum
> humano generi, superi, favistis abunde.
> ne foret hic igitur mortali semine cretus,
> ille deus faciendus erat.

> Caesar is a god in his own city—the man, distinguished in matters both military and civil, who was converted into a new star and a fiery comet not so much by his wars which finished in triumphs, his civil achievements or the renown which quickly followed, as by his own offspring. For there is no greater work among the acts of Caesar than the fact that he stands out as the father of this man. Is it really a greater achievement to have subdued the sea-girt Britons, to have led a victorious fleet across the seven-mouthed river of the papyrus-bearing Nile, to have added to the Roman people the rebellious Numidians, Libyan Juba and Pontus, swollen with pride from the name of Mithridates, to have celebrated some triumphs and to have earned many more, than to have given birth to so great a man? By his protection of all things you have, o gods above, shown favour in abundance to the human race. So that this

[4] See in particular Feeney (1991) 205–24, esp. 210–14; also Holleman (1969) 47–9; Due (1974) 86–8; Hardie (1997) 190–4; Salzman (1998) 330–2. The case for the politically subversive trajectory of the final part of *Metamorphoses* is made most forcefully by Holleman (1969).

[5] The text of Ovid's *Metamorphoses* is taken from Anderson (1977).

> one [Augustus] might not have been created from mortal seed, the other [Caesar] had to be made a god.

This is by no means uncomplicated panegyric. The emphasis on the relative importance of Augustus for Caesar's status (articulated here three times, at 15.746–50, 750–1, 752–8, and later again at 15.850–1, 852) has the effect of downgrading all of Caesar's military and civic achievements in a way which might be felt to compromise a strong conceptual link in Roman thought between mortal accomplishments and divine reward.[6] The remarkable sentiment at 15.760–1 only increases this sense of disjuncture. No agent is offered for *faciendus* at 15.761, and although it is Venus who emerges as the primary instigator of the apotheosis, working under the direction of Jupiter, the reader might well have in mind at this stage Octavian's own orchestration of Caesar's divinity. Either way, one is left with the strong impression that it is the agency of powerful supporters, gods and/or Octavian, rather than the merits of the individual, which leads to Caesar's apotheosis. A second area of tension surrounds the emphasis on Octavian and Julius Caesar's divine lineage to Venus. While the narrative works hard to press home the family connections,[7] this very emphasis might be felt to draw attention to its status as an imperial construct, and fissures in the rhetoric are exposed elsewhere in the episode.[8] Moreover, it is at least unfortunate that this episode proceeds directly after the story of Aesculapius' arrival into Rome (15.626–744): Aesculapius was welcomed as a god into Rome by both the people and the Senate—details conspicuous by their absence from the Caesar episode—and, as an individual who passed from divine to mortal and then back to divine state, Aesculapius provides an awkward reminder of the instability of godly status.[9]

While I would agree that the episode resists a straightforward reading of any kind, I judge the scholarly case for subversion to be too strong on two fronts. First, though it might seem a strangely anachronistic means of constructing praise, it should be recognised that other contemporary authors flatter Caesar in terms of the superiority of his

[6] The merits of the individual are likewise eclipsed in the deification stories of Hercules, Romulus and Hersilia in *Metamorphoses*; see Feeney (1991) 206, 208; Salzman (1998) 324–5; Tissol (2002) 328–34.

[7] Cf. 15.750 *sua progenies*; 751 *pater*; 758 *genuisse*; 775 *mei generis*; 776 *in me acui sceleratos cernitis enses*; 804 *Aeneaden*; 834–9, 844–5 *suique Caesaris*.

[8] Ovid offers a more down-to-earth assessment of Caesar when he refers to him as *mortali semine* (15.760), and Venus later bemoans the fact that Caesar represents the last of the Julian race (15.767 *quod de Dardanio solum mihi restat Iulo*).

[9] See Feeney (1991) 208–12; Salzman (1998) 319–20.

son.[10] Secondly, as will become clearer in the discussion later, I would contend that it is oversimplistic to view negatively Ovid's choice to advance a story of apotheosis achieved through divine agency.

In what follows, I argue that there were, by Ovid's day, different conceptual standpoints on offer with regard to apotheosis, in that it was open to the same level of creative enquiry as any other aspect of Roman religious practice in the late Republic and early empire.

ROMAN APOTHEOSIS: THE DYNAMICS OF A RELIGIOUS ENQUIRY BEFORE AND AFTER CAESAR

The first attested Roman author to deal with the concept of apotheosis is Ennius. Ennius translated into Latin Euhemerus of Messene's major work, *Sacred History*, which provided a rationalising account of deification as the result of exalted status achieved through earthly virtues and benefactions.[11] A strong connection between deification and earthly achievements resurfaces in an epigram that Ennius composed for Scipio Africanus, who is apparently the speaker of the fragment that survives (Vahlen 216):

> a sole exoriente supra Maeotis paludes
> nemo est qui factis aequiperare queat.
> si fas endo plagas caelestum ascendere cuiquam est,
> mi soli caeli maxima porta patet.

> From the rising sun beyond the marshes of Maeotis there is no one who would be able to match my deeds. If it is lawful for anyone to rise up to the regions of the gods, to me alone the supreme gate of heaven lies open.

[10] Cf. Prop. 4.6.59–60, where Caesar, speaking from his star, claims Octavian's exploits as proof of his own divinity ('*sum deus: est nostri sanguinis ista fides*'); cf. also Ov. *Fast*. 2.144 *caelestem fecit* . . . *ille patrem*, a sentiment which should be taken in both the literal and metaphorical senses. For a positive appraisal of Ovid's encomiastic style in this episode, see also Knox (1986) 75–8. More generally, Ovid's sentiment is consistent with a branch of scholarship that sees the downplaying of Caesar's earthly achievements as a conscious strategy in Augustus' own self-promotion; see Syme (1939) 317–18; Ramage (1985), esp. 240–1; for a forceful rejection of this view, however, see White (1988).

[11] Cf. Cic. *ND* 1.119 *quae ratio [i.e. apotheosis] maxime tractata ab Euhemero est, quem noster et interpretatus est et secutus praeter ceteros Ennius*, Vahlen 223–9. For the important place of Euhemerus in Hellenistic theories of apotheosis, see Bosworth (1999) 9–11.

Most famously perhaps, Ennius also treats the apotheosis of Romulus—the first Roman, in terms of mythological chronology, to be recognised as a god—in his epic *Annales*. Two fragments survive which relate to the deification of Rome's first king (Enn. *Ann.* 1. frr. 54, 110 Skutsch):

> unus erit quem tu tolles in caerula caeli
> templa
>
> There will be one whom you will raise into the heavenly temples of the sky
>
> Romulus in caelo cum **dis genitalibus** aevom
> degit
>
> Romulus lives his life in heaven with **the gods who bore him.**

Skutsch argues that, although fr. 110 suggests an apotheosis earned through divine origins, 'the Euhemerist and political moralist Ennius' would surely have used the episode to emphasise the role of earthly benefactions in procuring deification, in line with the intellectual thought-patterns of his other works.[12] But this is highly questionable. Granted that we cannot know what was contained in the missing parts of the epic, we should not overlook the strong evidence from *both* the remaining fragments, which present the consistent idea of apotheosis as an instance of divine homecoming and/or the work of divine agency: Romulus was already fated to be raised to heaven—at a time before he was rescued, at a council of the gods[13]—and was returned to the gods who gave him life in the first place (*genitalibus*).[14] Placing emphasis on divine connections and the agency of the gods in apotheosis may not, then, be a (subversive) penchant of Ovid's, but a perfectly respectful and traditional means of justifying an individual's translation to the skies, and one which might be felt to be particularly suitable for the epic genre which naturally foregrounds the majesty of the gods.[15] Even if Ennius did not retain a strong distinction between the roles of divine agency and earthly achievement

[12] See Skutsch (1985) 260. Skutsch's idea of an Euhemeristic colouring to Ennius' *Annales* is generally upheld by later scholars; see e.g. Cole (2006) 548; Myers (2009) 9.

[13] See Skutsch (1985) 202. This Ennian line is cited by Mars at Ov. *Met.* 14.814 and Ov. *Fast.* 2.487, and the contexts for both concur with this chronology.

[14] I agree with Skutsch (1985: 262) in interpreting *genitalibus* as *qui eum genuerunt*.

[15] This observation, famously articulated by Richard Heinze, holds true in general terms; see Heinze (1919) 10 with Hinds (1987) 99–114. It is also perfectly in line with the mythologising tendencies of *Metamorphoses* as a whole: see e.g. Tissol (2002) 328–9, 334: 'the fabulous incorporates all else into its domain—including history, politics, and current events'.

in procuring deification,[16] we should leave open the possibility that, across different works, Ennius adopted no fixed position on apotheosis and engaged rather in thematic and intellectual choices in this regard.

It is in this context of choice that we should read the first major treatment of apotheosis in a Roman setting, within Cicero's *Republic*, which was composed at a time when religion was becoming an area of academic enquiry in its own right.[17] More specifically, written against a backdrop of, and in response to, perceived disintegration of political order in the 50's BC, Cicero's work places virtue at the heart of good Roman government.[18] In fact, by adopting a strong connection between civic benefactions and status as a god, Cicero's work may even be designed to serve the (urgent) practical purpose of steering those contemporary leaders with divine ambitions towards greater service to the state.[19] Romulus, the primary precedent for a Roman ruler becoming a god, is therefore fashioned as a man whose wisdom and virtues towards the city lead directly to popular belief in his deification (*Rep.* 2.4–20).

But Cicero's adherence to a heaven won through earthly merit entails a subtle rejection of the view of apotheosis offered by Ennius in his *Annales* (*Tusc.* 1.28):

> ex hoc et nostrorum opinione 'Romulus in caelo cum diis agit aevum', ut famae adsentiens dixit Ennius

> Hence in the belief of our countrymen 'Romulus spends an age in heaven with the gods', as Ennius said in accordance with tradition.

As a near quotation of Ennius (*Ann.* 1. fr. 110 Skutsch), the key Ennian term *genitalibus* is conspicuous by its absence, as Cicero consciously erases the sense of divine dynasty found in his Ennian source to facilitate his more meritocratic vision of deification.[20] If indeed lost parts of

[16] See, for example, Cicero's sentiment that good deeds to the state demonstrate, in hindsight, divine birth (Cic. *Rep.* 2.4).

[17] See in general Chapter 4. For Euhemeristic apotheosis as a topic of debate during this period, see the divergent views expressed in Cic. *ND* 1.119 and 2.62.

[18] Cf. Cic. *Rep.* 1.1–13 with Zetzel (1995), esp. 1–3, 95.

[19] See especially the dream of Scipio (*Rep.* 6.9–29) with Zetzel (1995) 223–53; Cole (2006) 539–47. The notion of divine status as a reward for service to the Roman state also lies behind Cicero's (barbed) praise of Caesar in *Pro Marcello* (8, 27). For a similar notion in Varro, cf. *Ant.* fr. 20 Cardauns.

[20] Cicero evidently knew what he had missed out here: elsewhere, in a linguistic context, he readily acknowledges the exclusively Ennian collocation *di genitales* (*De Orat.* 3.154).

Ennius' *Annales* had advanced a view of deification achieved through earthly benefaction, it is difficult to explain why Cicero would have chosen not to quote from these parts and instead to resort to emendation of an otherwise inappropriate sentiment. Cicero's conscious appropriation of *Annales* here lends additional support to the view that Ennius was working with more than one thematic approach to deification across his works, on which a more single-minded Cicero felt compelled to make a stand.

If it is fair to say that Ennius and Cicero were operating within a context of choice with regard to apotheosis, then the dynamics of the negotiation would only have increased with the appearance of Caesar's comet and the popular interpretation that took hold concerning it. In the immediate aftermath of the comet, Octavian wasted no time in manufacturing a visual connection between Julius Caesar and astral imagery. He ensured that a star was added to the head of any statues of Caesar in Rome.[21] This visual association was soon also adopted on coinage from possibly as early as 44 BC.[22] The Senate duly ratified divine status for Caesar two years later, in 42 BC, with the full support of Octavian.[23]

None of these political or iconographic gestures, however, offers any prescription as to the perceived connection between Caesar and the stellar phenomenon, and it is to contemporary writers that we must turn to find possible interpretations. In particular, the comet provided for the first time a spectacular and high-profile *visual* referent for the concept of apotheosis. This seems to have prompted, again for the first time at Rome, some fervent negotiation concerning the actual process of translation from mortal to divine state.

It is now time to assess the wide-ranging ways in which contemporary Augustan writers engage with Caesar's comet. There is a vast amount of important scholarship on this topic—a topic that proves to be a conceptual minefield the further one ventures—on which I am

[21] Pliny tells us that Octavian added a star to a statue of Caesar in the Forum (*Nat.* 2.94). Similar insignia adorned the statues of Caesar in the Temple of Venus (Dio 45.7.1) and on the Capitoline (Serv. *Ecl.* 9.46). Servius (*Aen.* 8.681) goes as far as to say that all statues of Caesar were treated in this way. See further Scott (1941) 257–61; Ramage (1985) 238.

[22] See esp. Gurval (1997) 45–62, who shows effectively that, as with statuary, the symbol of the star took precedence over the comet in early post-Caesarian visual culture.

[23] See Weinstock (1971) 386–98.

grateful to draw.[24] By framing my own contribution on Augustan treatments of Caesar's comet within the context of an intellectual debate on apotheosis, I bring into the discussion a different and new range of source material, particularly from Ovid, by which I hope to demonstrate more clearly the vibrancy of a religious enquiry which operated within strict parameters of respect for Caesar's godhead.[25] Indeed, it is a notable feature of the Augustan accounts that, while there is dynamic debate concerning the 'mechanics' of apotheosis—the how and why—direct challenges to the underlying notion of Caesar's heavenly status are avoided.[26]

NEW SYMBOLISM FOR A COMET: CONTESTING POSITIVE RESONANCE

Before we can discuss the symbolism behind Caesar's comet, we must account for a complication that threatens to thwart any such enquiry from the outset: namely, there is an inconsistency in the terminology used to describe this stellar phenomenon, in that some contemporary commentators clearly refer to it as a star, rather than a comet. Recent scholarship has helped to make sense of this inconsistency,

[24] The complexities of this topic are most fully handled by Ramsey and Licht (1997). For other particularly valuable discussions of representations of Caesar's comet in contemporary literature and visual culture, see Scott (1941); Domenicucci (1996) 29–85; Gurval (1997). For earlier discussions of variant conceptions of the process of apotheosis, see Perret (1983); Le Boeuffle (1989) 125–9. On the general topic of aspects of apotheosis in ancient thought, the lectures of Franz Cumont (1922: 110–27, 148–69) are still illuminating.

[25] For references to Caesar's comet not covered in the following discussion, cf. Verg. *A.* 8.681 *patriumque . . . sidus*; Prop. 3.18.33–4. For further credible allusions to the comet, cf. Verg. *A.* 5.519–34 with e.g. West (1993) 9–13 (the positive omen of Acestes' flaming arrow); Hor. *Carm.* 1.12.46 *Iulium sidus* with e.g. Nisbet and Hubbard (1970) 162–3; Weinstock (1971) 378–9; Gurval (1997) 67 (a primary reference to Augustus, the 'star' of the Julian race, but an inescapable side allusion to the comet itself); Hor. *Carm.* 3.25.3–6 with Nisbet and Rudd (2004) 301–2.

[26] It is difficult to know what to make of Ov. *Am.* 3.8.51–2, where Ovid might be seen to offer criticism of the deification of mortals including Caesar; see e.g. Holleman (1971) 461–2. But the fantasy Golden Age context in which the sentiment is contained, as well as the company that the deified Caesar is allowed to keep (Romulus, Bacchus and Hercules), mitigate somewhat the negative force of Ovid's comment. I am grateful to Jim McKeown for his advice on these lines of Ovid.

which points principally to a strategic development in the identification of the phenomenon.[27] Leaving aside the general dispute among the ancients as to the identity of such stellar phenomena,[28] there are conscious reasons why, in the early stages, it might have seemed a risky tactic to have attached the symbol of a comet too firmly to notions of Caesar's divinity. As well as the predominantly negative omens associated with comets, there was the additional problem that a comet might connote a distinct lack of permanence, in that the Romans were undecided as to whether comets were fleeting fires or eternal celestial phenomena that merely slipped in and out of mortal view.[29] The symbol of a star, by contrast, would have seemed a much safer choice, as it connoted permanent status and had been associated with heaven-bound mortal leaders since the time of Alexander.[30] Only at a later stage, from the late 20's BC onwards, does the phenomenon of 44 BC become identified with a comet, at a time when the Augustan principate had moved onto a more secure political footing.[31] Most ambitiously, late in the Augustan age, Ovid attempts to harmonise these discordant traditions at *Met.* 15.848–50 (cited later), when he intimates that the fiery comet (*flammiferumque . . . crinem*) represents Caesar's transport to the heavens before finally metamorphosing into a permanent star (*stella*). In the following discussion of symbolism, therefore, I treat (early) references to Caesar's star and (later) references to Caesar's comet as indicative of the same celestial phenomenon that occurred in 44 BC.

The appearance of Caesar's comet opened up, essentially for the first time, an intellectual space for positive interpretations of such phenomena, and contemporary writers duly engage in this creative enquiry by giving voice to two different branches of thought, both of which were 'sponsored' by Octavian: first, in a reversal of the typical symbolism, that it represented a sign of positive rather than grievous change; secondly, and more radically, that it affirmed

[27] See esp. Gurval (1997); Ramsey and Licht (1997) 137–45.

[28] See Ramsey and Licht (1997) 139–40.

[29] For negative omens attached to comets, see Chapter 3. For Roman views on comets, see e.g. Gurval (1997) 42–3.

[30] For a history of astral immortality, see esp. Weinstock (1971) 370–84, who maintains that Caesar adopted the symbol of the star for himself during his lifetime: 'the comet then only served as a welcome confirmation of the existing symbol' (378).

[31] See Gurval (1997) 59–60; Ramsey and Licht (1997) 137–45.

Caesar's divinity.[32] Nor are we required to treat these two branches of enquiry as mutually exclusive, as the same author can evidently embrace both ideas across different works.[33]

If attaching positive significance to a comet was not entirely new before 44 BC, the young Octavian certainly appears to have given it political weight.[34] Pliny the Elder (*Nat.* 2.93–4), apparently drawing on Augustus' own written words, suggests that Octavian (privately) associated the comet positively with his own future career, a sentiment echoed by Baebius Macer (Serv. *Ecl.* 9.46), possibly a contemporary witness to the comet, who says more enigmatically that 'certain people' (*quidam*) associated the comet with a glorious career for Octavian. Virgil appears to take up this general cue in the first attested reference to Caesar's comet, delivered as part of a song recalled by the shepherd Lycidas (*Ecl.* 9.46–50, 42–40 BC):[35]

> 'Daphni, quid antiquos signorum suspicis ortus?
> ecce Dionaei processit Caesaris astrum,
> astrum quo segetes gauderent frugibus et quo
> duceret apricis in collibus uva colorem.
> insere, Daphni, piros: carpent tua poma nepotes.'

[32] I find unconvincing the view of Ramsey and Licht (1997: 14, 140–4) that 'rival', negative interpretations of Caesar's comet, entertained at the time of its appearance (Dio 45.7.1; Serv. *Ecl.* 9.46), continue into the Augustan age via Virgil's *Georgics* (1.487–8) and Tibullus (2.5.71). In these two cases, as Ramsey and Licht themselves acknowledge (1997: 14), the authors are careful not to associate their baleful comets specifically with Caesar's own. Indeed, Tibullus' comet is left entirely unspecified in terms of timing, and Virgil speaks of a different baleful phenomenon of multiple comets (*totiens . . . cometae*) within an unreal context of supernatural horror. I can see no reason to suppose that the positive association attached to Caesar's comet would have compelled all other comets to be treated (with hindsight) in the same manner. Rather, it is more natural to suggest that Augustan writers treated Caesar's comet as a positive exception to the general rule—in view of its exceptional brightness, daytime appearance and intensely politicised context—than to assert that all references to unspecified baleful comets in Augustan literature necessarily reflect badly on Caesar's own: see, for example, how Ovid, within the same episode, has no qualms in distinguishing the divine majesty of Caesar's comet from other comets that carry baleful connotations (*Met.* 15.787). The more directly controversial association between Caesar's comet and imminent warfare made by Calpurnius Siculus (*Ecl.* 1.77–83), a writer far removed from the Augustan era, is a quite different matter; for post-Augustan decline in respect for politicised astrology, see Chapter 11.

[33] See the differing interpretations offered by Virgil at *Ecl.* 9.46–50 and *A.* 8.681.

[34] For rare examples of, and scholarly caution towards, previous positively interpreted comets, see Ramsey and Licht (1997) 135–6.

[35] For the proposed dating of this poem, see Coleman (1977) 16. The text of Virgil is taken from Mynors (1969).

> 'Daphnis, why is it the ancient risings of constellations that you look upon? Look, the star of Caesar, born of Dione, has come forth, the star by which the fields may rejoice in their fruits and by which the grape may deepen its colour on the sunny hills. Graft your pears, Daphnis: your grandchildren will gather your fruit.'

Directing attention away from ancient celestial signs (*antiquos*), the singer exhorts Daphnis to look upon a new visible stellar phenomenon, emphatically categorised on two occasions as a star (*astrum*),[36] as ushering in an age of abundance in terms of both agricultural produce and, more implicitly, human fertility (via *nepotes*). In this way, Virgil takes Octavian's notion of a comet as symbolic of a positive future and produces a romanticised image of Caesar's star as the herald for a new age, one which recalls the Golden Age depicted in *Eclogue* 4.

This romanticised connection between the comet and Roman prosperity is not, however, taken up again by Augustan writers. The context in which the Virgilian sentiment is uttered helps to explain this silence. *Eclogue* 9 is set firmly in the aftermath of the Roman land confiscations of the early 40's BC and features a conversation between Lycidas and Moeris, who is leaving his farm now that it has been appropriated by a new settler (*Ecl.* 9.1–4). In this context, a celebration of the agricultural riches to be ushered in by Caesar's star reads now like a distinctly hollow prophecy. Though the identification of the 'star' with Caesar may be secure, Virgil subtly subjects to debate the precise nature of the positive significance behind the comet in the light of recent political upheavals.

In its place, it is a second and radically new interpretation, a connection between the comet and Caesar's divinity, which truly captures the Roman imagination. Moreover, within the parameters of this intellectual enquiry, we uncover a debate as to whether the comet was merely a sign of Caesar's apotheosis, or indeed the figure of Caesar himself. If we are to trust the testimony of later writers, both views appear to have been promoted by Octavian himself. Pliny the Elder purports to be quoting Augustus' own words when he writes (*Nat.* 2.94):[37]

> ipsis ludorum meorum diebus sidus crinitum per septem dies in regione caeli sub septemtrionibus est conspectum. id oriebatur circa undecimam

[36] For the use of *astrum* to denote a single star of significant size, cf. Isid. *Orig.* 3.60 *nam stella est quaelibet singularis . . . astra autem stellae grandes, ut Orion, Bootes. Sed haec nomina scriptores confundunt, et astra pro stellis et stellas pro sideribus ponunt.*

[37] The text of Pliny is taken from Mayhoff (1906).

horam diei clarumque et omnibus e terris conspicuum fuit. eo sidere **significari** vulgus credidit Caesaris animam inter deorum inmortalium numina receptam . . .

On the very days of my games a comet (lit. a hairy star) was seen for seven days in the northern region of the sky. It arose around the eleventh hour of the day and was bright and visible from all lands. The people believed that this comet **signified** that the soul of Caesar had been received among the dignified council of the immortal gods.

While the phraseology is ambiguous here, Augustus' use of *significari* suggests that the comet was first regarded as a sign of deification, rather than as a manifestation of the deceased dictator himself. By contrast Servius, who purports to be drawing on the contemporary witness of Baebius Macer, insists that Octavian was seeking a stronger connection between comet and Caesar (Serv. *Ecl.* 9.46):

Baebius Macer . . . dicit . . . quam quidam ad inlustrandam gloriam Caesaris iuvenis pertinere existimabant, **ipse animam patris sui esse voluit** . . .

Baebius Macer says that . . . while some people thought that it (the comet) had to do with the illumination of the young Caesar's glory, **Octavian himself wanted it to be the soul of his own father** . . .

Post-Augustan commentators continue to play out this debate,[38] but the focus here will be on the development of the debate within the Augustan age itself.

CROSSING FROM MORTAL TO DIVINE STATE: OVID AND THE DYNAMICS OF A NEW ENQUIRY

Before the time of Caesar's comet, the process of translation from mortal to divine status seems to have remained largely unconceptualised by Roman writers, along with related issues about what parts of one's constitution were mortal or eternal. The relevant fragments of Ennius, cited earlier, proffer no view. Cicero in his *Republic* follows Plato in

[38] For the comet as merely a sign of, or merely connected with, Caesar's divinity, cf. Plu. *Caes.* 69.4; Julius Obsequens 68; Serv. *Aen.* 6.790, 8.681. For the comet as Caesar himself, cf. Suet. *Iul.* 88 *creditumque est animam esse Caesaris in caelum recepti* ('and it was believed that it (the comet) was the soul of Caesar who had been received into heaven'). Dio 45.7.1 leans in both directions.

viewing the body as a mortal prison for an eternal soul, but he has little to say about the process of translation to heaven.[39] The appearance of a spectacular comet and its connection to Caesar's divinity, however, seems to have incited new avenues of enquiry in the minds of Roman thinkers, as an explanation was now invited as to the precise relationship between the recently deceased Caesar and the visible trail of the comet coursing through the sky. And it is Ovid who proves to be a focal point in any discussion of imaginative enquiry in this regard in the Augustan age.

First, let us return to Ovid's account of Caesar's apotheosis in *Metamorphoses*, concentrating on his handling of the actual process of translation from mortal to divine state (*Met.* 15.840–51):

'hanc animam interea caeso de corpore raptam
fac iubar, ut semper Capitolia nostra forumque
divus ab excelsa prospectet Iulius aede.'
vix ea fatus erat, media cum sede senatus
constitit alma Venus nulli cernenda suique
Caesaris eripuit membris nec in aëra solvi
passa recentem animam caelestibus intulit astris,
dumque tulit, lumen capere atque ignescere sensit
emisitque sinu: luna volat altius illa
flammiferumque trahens spatioso limite crinem
stella micat natique videns bene facta fatetur
esse suis maiora et vinci gaudet ab illo.

[Jupiter addressing Venus] 'In the meantime, snatch Caesar's soul from his slaughtered body and lend it a heavenly radiance, so that Iulius, as a god, may always look down from his exalted temple onto our own Capitol and forum.' Scarcely had he uttered these words when kindly Venus set down in the middle of the seats in the senate, unseen by anyone, and snatched away from the limbs of her own Caesar the fresh soul, not allowing it to be released into the air, and bore it to the heavenly stars; and while she was carrying it, she felt it take up light and catch fire, and sent it forth from her lap: it flew higher than the moon itself, trailing flame-bearing hair on its wide-ranging path, and shone as a star; and seeing the achievements of his son he confessed that these were greater than his own and rejoiced at being surpassed by his son.

[39] For the body as mortal prison for an immortal soul, cf. Cic. *Rep.* 6.14 with e.g. Plat. *Phd.* 67cd; *Phdr.* 250c; *Gorg.* 492e–493a; see also Zetzel (1995) 231. In terms of the process of translation, Cicero speaks only in vague terms of the soul entering the heavens; cf. *Rep.* 6.24 *cum Romuli animus haec ipsa in templa penetravit.*

Ovid presents a consistent image here of a soul detachable from a material mortal frame, and it is Caesar's soul—seemingly distinct in shape from its former body, and small enough to be carried in Venus' lap—which is transported aloft and, upon ignition, accounts for the appearance of the comet. This is consistent with the Ciceronian/Platonic view of the relationship between body and soul, an idea which appears to have become, by Ovid's day, a popular philosophical commonplace.[40]

In the light of the central thematic charge of *Metamorphoses*, it should come as no surprise that Ovid pays particular attention to the process of change from one state to another. What is perhaps more surprising is the way in which Ovid, through the medium of the other apotheosis episodes in the poem, appears to exploit thematic potential raised by the visibility of Caesar's comet to offer a range of divergent (and no doubt playful) views on the process of migration.

The first description of the process of migration in the epic occurs with Hercules, the archetypal mortal benefactor to be converted into a god (*Met*. 9.262–72):

> interea, quodcumque fuit populabile flammae,
> Mulciber abstulerat, nec cognoscenda remansit
> Herculis effigies, nec quicquam ab imagine ductum
> matris habet, tantumque Iovis vestigia servat;
> utque novus serpens posita cum pelle senecta
> luxuriare solet squamaque nitere recenti,
> sic, ubi mortales Tirynthius exuit artus,
> parte sui meliore viget maiorque videri
> coepit et augusta fieri gravitate verendus.
> quem pater omnipotens inter cava nubila raptum
> quadriiugo curru radiantibus intulit astris.

> In the meantime, Vulcan had stolen away whatever part could be devastated by the flame, nor did a recognisable resemblance of Hercules remain: he did not now have anything drawn from the image of his mother, and only vestiges of Jupiter remain. And just as when a snake, having laid aside its old skin, is accustomed to abound anew, resplendent with its fresh scales, so, when the

[40] Cicero is most direct about the difference in 'shape' between body and soul at *Rep.* 6.26, drawing on the suggestions in e.g. Plat. *Phd.* 115c–e; *Lg.* 12.959ab. For popular articulations of this general sentiment among Augustan writers, cf. Verg. *A.* 6.703–51, esp. 719–21, 730–2; Prop. 3.18.31–2; Ov. *Pont.* 4.13.25–6; specifically in relation to Caesar, cf. V. Max. 4.5.6; similarly of Augustus' perceived apotheosis, cf. Germ. 558–60 (discussed in Chapter 8).

Tirynthian shed his mortal limbs, he gained vigour in the better part of him and began to appear greater and to become venerable in his sacred gravity. His almighty father stole him away amid the hollow clouds and bore him to the shining stars in his chariot drawn by four horses.

In comparison with the narrative of Caesar, it is more difficult to decipher what precisely is involved in Hercules' translation from mortal to divine state. Like Caesar, deification involves separation from mortal limbs (268 *mortales . . . artus*), but are we justified in interpreting Hercules' 'better part' (269 *parte . . . meliore*) as his soul? The context speaks against this idea. First, there is no specific reference to *animus*, and Jupiter has earlier spoken of Hercules' eternal part as something specifically fashioned by him (252–6), a sentiment which alludes more readily to Hercules' status as the product of both mortal and divine parentage. Secondly, and in apparent opposition to 263–4 (*nec cognoscenda remansit Herculis effigies*), the metaphor of the snake sloughing its skin encourages us to think in terms of two similar forms of *corporeal* covering, with the inner layer judged to be fresher and brighter than the outer, a motif which is consistent with the popular Greco-Roman tendency to anthropomorphise their gods as more resplendent versions of human beings.[41] We are encouraged, then, to view the migrating entity as something which retains some semblance of earthly features, somewhat different from the (shapeless) soul that Venus held to her bosom in the case of Caesar. Granted that Caesar and Hercules represent different types of divine 'offspring', a comparison of two stories of apotheosis already reveals our poet's playing with different conceptions of what exactly is carried to heaven for deification, and what exactly is left behind on earth.

The next apotheosis story, concerning Aeneas, strikes a similarly corporeal, anthropomorphic note to that of Hercules (*Met.* 14.600–8):

hunc iubet Aeneae, quaecumque obnoxia morti,
abluere et tacito deferre sub aequora cursu.
corniger exsequitur Veneris mandata suisque,
quidquid in Aenea fuerat mortale, repurgat
et respersit aquis: pars optima restitit illi.
lustratum genetrix divino corpus odore
unxit et ambrosia cum dulci nectare mixta

[41] For Roman anthropomorphism of their gods, see e.g. Levene (2012) 64–6. For corporeal conceptions of apotheosis, see esp. Perret (1983), as well as the discussion later.

> contigit os fecitque deum; quem turba Quirini
> nuncupat Indigetem temploque arisque recepit.

> She [Venus] ordered him [Numicius] to wash away from Aeneas whatever parts were subject to mortality and to carry him beneath the surface of the river in a silent course. The horned one carried out Venus' instructions, and he purged out and sprinkled over with his own waters whatever had been mortal within Aeneas: only the best part of him remained. His mother anointed his purified body with a divine fragrance, touched his face with ambrosia mixed with sweet nectar, and made him a god. The people of Quirinus call him Indiges by name and received him with a temple and altars.

If Hercules' mortal part had been purged by fire, so Aeneas' is purged by the water of the river Numicius to leave 'the best part' (604 *pars optima*).[42] There is no detail of translation to heaven here, as Venus appears to fashion Aeneas as a god while he is still on earth. But as with Hercules, it is unclear whether Venus is working with Aeneas' soul or, more likely, with a more human-like corporeal entity which retains at least the semblance of a body and face (cf. 605 *corpus*, 607 *os*).

The deification of Romulus, by contrast, presents us with a quite different idea of the process of translation (*Met.* 14.824–8):

> abstulit Iliaden: corpus mortale per auras
> dilapsum tenues, ceu lata plumbea funda
> missa solet medio glans intabescere caelo;
> pulchra subit facies et pulvinaribus altis
> dignior est qualis trabeati forma Quirini.

> He [Mars] stole away the son of Ilia: his mortal body dispersed through the thin breezes, just like a leaden bullet sent forth from a wide sling is accustomed to melt away in the midst of the sky; a beautiful countenance appeared in replacement, and it was more worthy of exalted couches, just like the form of robed Quirinus.

Although he is again playing with familiar ideas of a purer divine form bereft of mortal covering, this is the first occasion on which Ovid suggests that the mortal body itself starts a journey towards heaven before disintegrating in mid air. To lend scientific credibility to this notion, Ovid provides the comparison of the leaden bullet which, when

[42] For ritual purification by water, see *Met.* 13.951–3 with Hopkinson (2000) ad loc.; Verg. *A.* 6.741–2. For water deities who undergo the same process of having their mortal part washed away by water, cf. *Met.* 4.539–42 (Ino and Melicerte), 13.949–55 (Glaucus).

propelled from a sling, was widely believed to melt away owing to its speed of movement through the air.[43] In the light of its inclusion within both scientific discourse and high hexameter poetry, I do not subscribe to the view that the simile here produces incongruity or undermines the gravity of Romulus' deification. On the contrary, it can be read, retrospectively, as an attempt to anchor this mythical event to the reality of Rome and the comet of Caesar. So great is this episode's gravitational pull towards the apotheosis of Caesar that the more popular version of this part of the story, that Romulus sought the heavens in the chariot of his father Mars, has been squeezed out to make room for a visible molten mid-air display.[44] The phenomenon of the leaden bullet, then, offers with hindsight an explanation of the comet in terms of the process by which Caesar's body burns away during his journey to heaven.

The reception into heaven of Romulus' wife, Hersilia, narrated shortly after this event, maintains the focus on the idea that the disintegration of mortal elements occurs at a relatively late stage in the process of apotheosis (*Met.* 14.845–51):

> nec mora, Romuleos cum virgine Thaumantea
> ingreditur colles: ibi sidus ab aethere lapsum
> decidit in terras, a cuius lumine flagrans
> Hersilie crinis cum sidere cessit in auras.
> hanc manibus notis Romanae conditor urbis
> excipit et priscum pariter cum corpore nomen
> mutat Horamque vocat, quae nunc dea iuncta Quirino est.

> Without delay, she [Hersilia] approached the hill of Romulus with the virgin daughter of Thaumas [Iris]: there a star gliding down from heaven fell upon the earth, and Hersilia, her hair aflame from its light, departed into the air along with the star. The founder of the city of Rome receives her with his famous hands and, at the same time, changes her name along with her body and calls her Hour, and now as a goddess she is joined to Quirinus.

In this story, Hersilia's earthly frame is only changed when she reaches heaven, and a visible sign of her migration is offered in the form of

[43] For the popular belief, which can be traced back at least to Aristotle, cf. Arist. *Cael.* 2.7; Lucr. 6.177–9, 6.306–8; Verg. *A.* 9.588–9; Ov. *Met.* 2.726–9; Luc. 7.513; Sen. *NQ* 2.57.2; Stat. *Theb.* 10.533.

[44] For Mars' traditional role in facilitating Romulus' translation to the heavens, cf. Hor. *Carm.* 3.3.15–16; Ov. *Fast.* 2.496 and the implications of Enn. *Ann.* 1. fr. 54 Skutsch. In our episode, Mars descends to earth in his chariot and picks up Romulus' body, before a quasi-comet narrative takes over (*Met.* 14.818–26).

the flaming hair of her body. This image betrays the clearest influence from Caesar's comet, as comets are conceptually connected with hair.[45] Retrospectively, then, Hersilia, a mortal with flaming hair catching a ride on a star, provides yet another possible way of conceptualising the relationship between Caesar and his comet.[46]

To summarise, by the time that he has narrated the apotheosis of Caesar in *Metamorphoses* 15, Ovid has already set up a number of choices for the reader as to how a (spectacularly visible) translation from mortal to divine status might be interpreted, to which the actual interpretation offered in the Caesarian narrative provides additional choice rather than definitive clarification. Do we accept the interpretation offered in the episode itself, that we are witnessing the light from a shapeless soul after its release from the mortal frame? Or do we revert to previous stories of apotheosis in the epic to interpret the comet's light as the actual moment of separation of physical body from immortal soul, or as the physical hair of the mortal frame as it is carried along by a heavenly being? In a similar context of choice, the depiction of Caesar's soul as a shapeless entity jostles for position in the poem with ideas about the corporeality of a soul that retains at least a semblance of its earthly features.

Ovid's contemporaneous religious poem, *Fasti*, takes this negotiation still further (*Fast.* 3.697–702):

> praeteriturus eram gladios in principe fixos,
> cum sic a castis Vesta locuta focis:
> 'ne dubita meminisse: meus fuit ille sacerdos;
> sacrilegae telis me petiere manus.
> ipsa virum rapui simulacraque nuda reliqui:
> quae cecidit ferro, Caesaris umbra fuit.'

[45] For *crinis* used to denote stellar phenomena (shooting stars and comets), cf. e.g. Verg. *A.* 5.528; Man. 1.835–9; *TLL* 4.1204.70–1205.16. For etymological connections between *cometes* and *coma/crinis*, cf. e.g. Hyg. *Astr.* 2.21; Plin. *Nat.* 2.89; Maltby (1991) 142 s.v.

[46] Related to this conception of apotheosis are depictions that appear to treat the deified Caesar as separate from, and dwelling on, a star; cf. Prop. 4.6.59–60 *at pater Idalio miratur Caesar ab astro:/'sum deus; est nostri sanguinis ista fides'*. The depiction of a corporeal Roman individual being transported to heaven may be influenced by emerging imperial iconography for apotheosis which depicts members of the ruling family carried aloft by chariots, birds and other winged couriers; see, for example, the so-called Belvedere Altar (12–2 BC), which depicts (probably) Caesar being carried to heaven on a horse-drawn chariot, with Zanker (1988) 220–1; Kleiner (1992) 102–3. See further Kleiner (1992) 285–8, 311–12; Beard and Henderson (1998).

> I was about to pass over in silence the swords that fixed themselves in the body of the princeps, when Vesta thus spoke to me from her chaste hearth: 'Do not hesitate to remember: my priest he was, and sacrilegious hands attacked me with their weapons. I myself stole away the man and left a barren likeness: that which fell by the sword was the shadow of Caesar.'

One cannot deny an element of humour here, as Vesta, possibly reacting against (and embarrassed by) the centrality of Venus in the version of events offered in *Metamorphoses*, presents an egocentric account which puts herself at the forefront of sympathy and activity on behalf of her own priest.[47] But whether or not we regard her as a reliable authority, it is noteworthy that Vesta puts forward yet another conception of apotheosis, one which specifically creates two different Caesars, the real one transported wholesale to heaven, and an empty image (*simulacra*) or shadow (*umbra*), which is left on earth for the assassins.[48]

A final contribution to the enquiry on the process of migration comes, I would suggest, from an unlikely source. At first glance, the careers of Caesar and the mythical hunter Hippolytus are worlds apart. And yet Ovid's version of Hippolytus' death, narrated in the same book as his treatment of Caesar (*Met.* 15.492–546), brings the two individuals into closer proximity than one might think: both are undone by the trickery of wicked individuals, and both suffer death by impalement before being converted into a god by means of divine assistance.[49] This unique alignment makes more interesting the central novelty involved in Ovid's handling of Hippolytus: that it is the protagonist himself who offers the account of both his death and his transformation into a god. Motivated by more than a simple penchant for poetic innovation, I would venture the view that Ovid is seeking to provide a further perspective on the Caesar phenomenon by forwarding the notion that

[47] Vesta may be taking objection to the fact that Venus speaks for her directly (*Met.* 15.777–8). For implicit direction from Ovid to read *Fasti* after *Metamorphoses*, see e.g. Green (2004) 28–9. For the use of words of memory, including *memini*, to trope textual reminiscences in Ovid (here = 'do not hesitate to remember (viz. correctly, regardless of what Venus has said)'), see Hinds (1998) 3–4.

[48] This idea may have been inspired by the curious lines in Homer which suggest that the real Herakles dwells in heaven, while only his image travels to earth (*Od.* 11.601–4).

[49] For the trickery and wickedness involved in their deaths, cf. 15.498 *sceleratae fraude novercae* (Hippolytus) with 15.766 *fraude*, 15.776 *sceleratos . . . enses*. The detail of Hippolytus' impalement by a branch or tree trunk (15.525 *nervos in stipe teneri*) appears to be an Ovidian contribution to the myth which is subsequently taken up by Seneca (*Phd.* 1098–9); see further Segal (1984) 323. Hippolytus receives godly help from both Aesculapius and Diana in his conversion to divine status (15.533–46).

there is a continuum of consciousness from mortal to divine form, even in cases where the body is badly damaged. Through the mouth of Hippolytus, the reader can be comforted—if not the internal listener, Egeria (15.493–6, 547–8)—by the fact that the divinised Caesar shining from the skies shares the same memories as the defiled mortal on earth.

ACCOUNTING FOR APOTHEOSIS: A FAMILIAR ISSUE REVISITED

Finally, we come back to the issue that provoked debate even before the appearance of Caesar's comet, namely the justification for the apotheosis of mortals. I argued earlier that Ennius and Cicero play out a debate as to whether apotheosis is to be seen principally as the result of divine intervention/nepotism or the reward for earthly benefactions: dynasty versus meritocracy. This enquiry shows no signs of abating in the Augustan age.

Most interestingly, and in spite of some modern scholarly disquiet, it must be acknowledged that Ovid's depiction of a Caesar translated to the heavens through divine agency is, if anything, the more popular interpretation in the Augustan age.[50] The earliest articulation of this general idea comes, surprisingly, from the introduction to the technical prose treatise of Vitruvius in the 20's BC, within an expression of the author's loyalty from one generation of Caesar to another (1 *praef.* 2):

> . . . primum parenti tuo de eo fueram notus et eius virtutis studiosus. cum autem concilium caelestium in sedibus inmortalitatis eum dedicavisset et imperium parentis in tuam potestatem transtulisset, idem studium meum in eius memoria permanens in te contulit favorem.

> I had first been known to your father [Caesar] and had been devoted to his virtue. But when a council of the gods had consecrated him in the abode of the immortals and had transferred the father's power to

[50] It is impossible to know what to make of Varro's now fragmentary *De Gente Populi Romani* (43–42 BC), a work which deals with the divine honours received by legendary kings and may have included the deification of Romulus (see Fraccaro 1907)). It may be that this work, written shortly after Caesar's comet, implicitly endorses a meritocratic impression of Caesar's deification through association; see Taylor (1934); Horsfall (1972) 124–5. But this is a matter of debate, as existing fragments may point towards a more problematic and sobering view of apotheosis from Varro; see Baier (1999).

you [Augustus], that same enthusiasm persisted, in memory of him, and brought my support to you.

Though Vitruvius acknowledges Caesar's virtue, it is noticeable that he keeps this sentiment separate from Caesar's apotheosis, opting instead to foreground the role of divine agency. If, as I have argued earlier, Vitruvius is a man genuinely seeking to align himself with elite imperial sensibilities, then our author's advancement of divine intervention over (straightforward) meritocratic sentiment in the case of Caesar becomes important evidence of Augustus' support for the former. Evidently, this idea is not the preserve of the fabulous world of epic poetry. A further articulation of this idea can be found in Ovid's religious poem, at the end of a discussion about the development of the Roman calendar (*Fast.* 3.155–60):

> sed tamen errabant etiam nunc tempora, donec
> Caesaris in multis haec quoque cura fuit.
> non haec ille deus tantaeque propaginis auctor
> credidit officiis esse minora suis,
> promissumque sibi voluit praenoscere caelum
> nec deus ignotas hospes inire domos.

> Nevertheless the calendar [lit. days] was even now still in error, until this also became one of Caesar's many cares. Not this care did that god, author of so great a progeny, believe to be beneath his duties, and he wanted to know in advance the heaven promised to him, nor did the god enter into unknown homes as a stranger.

Ovid lays it on thick here with an equally strong sense of Caesar's divine origins and celestial trajectory: he is already a god (referred to twice as a *deus* on earth) for whom heaven has already been promised.[51] In the light of Caesar's major contribution to the calendar, a meritocratic sentiment could easily have been advanced to account for his apotheosis, but Ovid chooses to emphasise divine dynastic connections.

By contrast, a straightforward interpretation of Caesar's apotheosis as earned through earthly benefactions is nowhere to be found in the Augustan age. Ovid, as we have seen (esp. *Met.* 15.752–8), initially contemplates a virtue-based approach—and referring to Augustus as the pinnacle of Caesar's 'achievements' is, to say the least, an ironic touch—only to pursue the alternative, divine-centric explanation for

[51] For the notion that Caesar is heaven-sent and destined to return to the gods, cf. also Man. 4.57–62.

apotheosis.[52] In similar vein, a little later, in the Tiberian age, Valerius Maximus' meritocratic interpretation of Caesar's apotheosis contends with the view that his deification amounts to a special case of homecoming for a divine soul (6.9.15, 4.5.6):[53]

> C. autem Caesar, cuius **virtutes** aditum sibi in caelum instruxerunt . . .
>
> But Caesar, whose **virtues** equipped him for an approach to heaven . . .
>
> quam praecipuam in C. quoque Caesare fuisse et saepe numero apparuit et ultimus eius dies significavit: compluribus enim parricidarum violatus mucronibus, inter ipsum illud tempus, quo **divinus spiritus** mortali discernebatur a corpore, ne tribus quidem et viginti volneribus quin verecundiae obsequeretur absterreri potuit, si quidem utraque togam manu demisit, ut inferior pars corporis tecta conlaberetur. **in hunc modum non homines expirant, sed di immortales sedes suas repetunt.**
>
> This quality [sc. dignity] was especially showing in Julius Caesar; it was often evident and the last day of his life signified it. For violated by the many blades of his assassins, at that very moment when his **divine spirit** was being separated from his mortal body, not even twenty-three wounds could deter him from observing modesty: indeed, he let his toga fall with both hands so that the lower part of his body was covered as he collapsed. **In this way, it is not the case that men expire, but rather that immortal gods seek again their own abodes.**[54]

To conclude, the special phenomenon of Caesar's comet, burning brightly for several days in the sky, provided the impetus for a branch of religious enquiry which called out to intellectual creativity. And yet this enquiry was played out within agreed parameters of respect for the underlying notion of the deceased ruler's astral immortality. Perhaps the best testament to this overall mindset is offered some years after the death of Augustus by Valerius Maximus, whose negotiations on why exactly Caesar was deified (see earlier) have no bearing on the 'fact' of his apotheosis.[55] In this way, contemporary writers lend legitimacy to the Emperor's divine heritage, as well as the power of the stars, in a manner which eschews politically sensitive aspects of astrology.

[52] In similar fashion, Manilius combines the motifs of meritocracy and divine homecoming/nepotism in his introductory address to the Emperor at 1.7–10.

[53] The text of Valerius Maximus is taken from Briscoe (1998).

[54] For the motif of Caesar as heaven-sent in Valerius Maximus, see further Wardle (1997) 336–42.

[55] Cf. e.g. V. Max. 3.2.19 *siderum clarum decus, divum Iulium*; 9.15.1 *divinae Caesaris vires. caelo receptus est.*

10

Turning on the Practitioner

Propertius Meets the Charlatan

One consistent threat to the credibility of astrology, frequently levelled by its philosophical detractors, is the fact that some predictions have turned out to be inaccurate; these become, in the hands of the sceptic, powerful tools in the quest to invalidate the science itself. In tackling this issue directly at the beginning of his *Mathesis*, Firmicus Maternus (4th century AD) draws a sharp distinction between the science, which is regarded as truthful, and the particular practitioner, who may be ignorant or fraudulent (*Math.* 1.3.6, 8):[1]

> si vero erraticis responsionibus is, qui mathesin se scire profitebatur, non potuit implere, quae dixerat, et responsa eius mendacia promissionis inmaculant, imperitia illius temeritasque pulsanda est et fallax infamanda promissio; omni enim malo dignus est, quod per illum stet, ut praeclarae artis studium falsis responsorum mendaciis obumbretur.

> If indeed, through erratic responses, the practitioner who claims to know the science of astrology is not able to substantiate his statements, his responses are stained by the falsehood of his forecast—the ignorance and foolhardiness of that man should be driven away and his deceptive forecast brought into disrepute. For he is deserving of every punishment, because it is through him that study of this noble art is overshadowed by lying responses.

[1] The text of Firmicus Maternus is taken from Kroll and Skutsch (1897–1913). Ptolemy in his *Tetrabiblos* offers a similar distinction (1.2.6–7), but the force of this is compromised by his belief that *all* astrologers are to some extent imperfect; Firmicus, by contrast, maintains a clear-cut distinction between competent and inaccurate practitioners (*Math.* 1.3).

> si vero interroganti ei imperitus professor falsa respondit, non mathesis, sed hominis fallax ac temeraria notetur inscientia.
>
> But if an unskilled practitioner has given false responses to his client, then let it not be the science that is blamed, but the deceitful and foolhardy ignorance of the man involved.

The emphasis on ignorance and deceit recalls most readily the sort of amateur astrologer that could be hired on the streets of Rome.[2] Though articulated most directly in Firmicus Maternus, this distinction between noble art and suspect practitioner can be detected much earlier in Roman history, during the early empire, at a time when authors are seeking safe but attractive means of handling the topic of astrology.[3] And it is Propertius who takes up this challenge.

HOROS IN PROPERTIUS 4.1B: READING AN ASTROLOGICAL CONFIDENCE TRICKSTER

Propertius' depiction of Horos, who appears in the second of a pair of poems that open Propertius' fourth book, is probably the most sustained treatment of astrology in Augustan literature.[4] Propertius opens the book (1A) by acting as guide for a stranger to Rome, as he points out the sites of the great city and compares them to the state of the area in its early rural days (1–54). This leads on to a specific programmatic pledge to sing an aetiological poem about the city and become, in effect, the Roman equivalent of Callimachus, the author of *Aetia* (1A.61–4):[5]

[2] See Chapter 3. For the emphasis on ignorance and deceit from these peddlers, cf. Hor. *Sat.* 1.6.113–14 (*fallacem Circum*); for low-life diviners at Rome, cf. also Juv. 6.582–91.

[3] See Barton (1994a) 50–1, esp. 50 n. 117, where specific examples are adduced from Pliny the Elder and Apuleius. But Barton does not offer any detailed example of this phenomenon before the late 1st century AD.

[4] Originally transmitted in the manuscripts as a single poem, modern scholars tend to break this up into a pair of poems, typically labelled 1A and 1B: Prop. 4.1 is felt to be neither coherent enough to be treated as one continuous poem nor distinct enough in subject matter to be treated as two entirely separate entities. For the differing approaches, see Hutchinson (2006) 61; Heyworth (2007b) 424–5.

[5] The text of Propertius is highly problematic. I follow Hutchinson (2006) for the text of Prop. 4.1 unless otherwise indicated (although it is printed as one poem in this edition).

Ennius hirsuta cingat sua dicta corona:
mi folia ex hedera porrige, Bacche, tua,
ut nostris tumefacta superbiat Vmbria libris,
Vmbria Romani patria Callimachi!

Let Ennius surround his verses with a shaggy crown: O Bacchus, stretch out for me leaves from your ivy so that Umbria may swell with pride at my books, Umbria, the fatherland of the Roman Callimachus!

Poem 1B, however, abruptly tempers this sense of poetic enthusiasm by bringing on to the scene an astrologer named Horos, whose central message, despite much long-windedness, is that Apollo does not support Propertius' intention to leave the sphere of love elegy to embark on this grander aetiological project. Propertius provides no direct response to Horos, as the poem ends straight after Horos' speech, but the effect of the astrologer's warning may be felt implicitly from Book 4 as a whole which, depending on one's point of view, either completely abandons or seriously compromises on the pledge made in 1A to compose a Roman aetiological poem.[6] Consequently, previous scholarship has typically analysed Horos within the wider context of Book 4, seeking to ascertain his contribution to the literary programme of the final book as a whole.

Given the focus of this study, however, I wish to treat Horos in isolation and analyse in some detail the representation of an astrologer who apparently operates at Rome and in Propertius' own (Augustan) age.[7] In what follows, I argue that Propertius finds a safe and amusing means of handling astrology by portraying Horos as a charlatan—a confidence trickster whose boasts of astrological prowess ultimately ring hollow—while being careful to avoid tarnishing the reputation of the art itself.

At the start of 4.1B, an individual attracts the attention of Propertius by using his name and directly contradicting the sense of optimism that the poet has just expressed in his new aetiological project (71–4):

quo ruis imprudens †vage† dicere fata, Properti?
non sunt a dextro condita fila colo.

[6] For the view that Propertius abandons his initial pledge, see e.g. Stahl (1985) 265–305; for the (more popular) viewpoint that Book 4 amounts to some sort of fusion between the erotic elegiac and the aetiological, see esp. Debrohun (2003). For a useful history of scholarship on this issue, see Stahl (1985) 265–9.

[7] Propertius' fourth book is typically dated to *c.* 15 BC and no earlier than 16 BC; see e.g. Hutchinson (2006) 2–7.

accersis lacrimas cantans; aversus Apollo.
poscis ab invita verba pigenda lyra.

To where are you rushing . . . ignorant of how to speak your destiny, Propertius? The threads have not been spun from a favourable distaff. You are summoning forth tears by your singing: Apollo is in opposition. You demand from an unwilling lyre words which would have to be regretted.

Whoever this individual is (unknown at this stage), he/she has clearly been privy to the final part of Propertius' conversation with the stranger at 1A.67–70: Propertius' metaphorical expressions of speed (70 *sudet*) and intended destination (70 *has . . . ad metas*) are met immediately with scepticism, as the unknown individual acknowledges the pace and questions the destination (71); *non a dextro . . . colo* (72) directly negates the poet's wish for a favourable omen for his undertaking (68 *dextera . . . avis*) and Horos redirects Propertius' use of *canto* (73, cf. 68) towards a negative context.

Having attracted the attention of the poet in this way, the individual turns to himself and his craft (75–80):

certa feram certis auctoribus, aut ego vates
nescius aerata signa movere pila.
me creat Archytae suboles, Babylonius Horops,
Horon, et a proavo ducta Conone domus.
di mihi sunt testes non degenerasse propinquos,
inque meis libris nil prius esse fide.

Sure things will I reveal from sure authorities, or else I am a prophet who does not know how to move the constellations on the bronze ball. The offspring of Archytas, the Babylonian Horops bore me—Horos by name—and my family goes back to Conon, my ancestor. The gods are my witness that I have not disgraced my relatives, and that in my books there is nothing before the truth.

The individual, then, tells us that he is an astrologer named Horos.[8] What are we to make of him? A surface reading of these lines suggests

[8] The name of this fictitious character is aptly chosen, as it evokes simultaneously: Horos (Ὧρος), the Egyptian deity equivalent of the prophetic Apollo; ὅρος ('limit'), reflecting Horos' advice to the poet not to go beyond his generic remit; ὥρα ('season, year'), etymologically linked to 'horoscope'; see further Debrohun (2003) 19–21.

a figure of considerable authority and heritage.[9] He is from Babylonian stock, a nation regularly associated with astrology, and he counts as ancestors names that one would associate with two famous Greek astronomers, Archytas and Conon (77–8).[10] He has, moreover, lived up to this impressive ancestry in his own time, in that he has not brought any disgrace on his present family (79). Bolstered by this personal history, Horos can claim confidently that he will speak certainty in which one can trust: here and elsewhere in his speech, *certus* and *fides* are rammed home as hallmarks of his particular trade (75, 80; cf. also 92, 98).

Granted that the charismatic listing of one's credentials might have been a requisite part of any initiation of 'trade' between a diviner and his clients,[11] we ought to be wary of an individual who is so obsessed with singing his own praises: in addition to the emphasis on truth and certainty, note also the inflated self-stylisation in *vates* (75) and the consistent use of first person pronouns/adjectives: *ego* (75, 89, 99), *mihi . . . meis* (79–80), *meam* (92). In fact, as Stahl duly notes (1985: 270), Horos spends 72 of the 80 verses of his speech on topics related to his credentials, and only the last four lines (147–50) contain anything potentially astrological (for which see later).

This obsession, one might say desperation, on the part of the speaker points in the direction of the astrological charlatan, the sort of peddling diviner who might have courted business in the streets of Rome.[12] The dramatic context for the two parts of poem 4.1 lends support to this theory: Propertius has been walking around Rome acting as a guide for his guest, exactly the sort of trip during which one might

[9] For the view that Horos is a serious, creditable astrologer, see esp. Montanari Caldini (1979); also Sandbach (1962) 267; Macleod (1976); Keyser (1992); Debrohun (2003) 73; Welch (2005) 32–3.

[10] Archytas was a famous mathematician and Pythagorean of the 4th century BC; Conon was the court astrologer of Ptolemy Euergetes and Berenice who famously claimed a place for the Lock of Berenice among the stars close to Leo; see further Hutchinson (2006) 75.

[11] See, with respect to the Greek world, Flower (2008) 70–1, 92–100.

[12] Stahl (1985: 270) suggests in passing that Horos resembles 'a cheap seer in the streets of Rome', but he does not pursue this line of enquiry. Ripat (2011: 130), who considers Horos to be 'a chimaera of name and lineage', suggests that Propertius might be commenting more generally on the fanciful Eastern credentials that such figures deceitfully claim. For earlier (largely unsubstantiated) assessments of Horos as a ludicrous, unconvincing figure, see Warden (1980) 102; esp. Montanari Caldini (1979) 13–14. Courtney (1969) 73–80 and Hutchinson (2006) 73–86 *passim* offer useful discussions of the absence of a proper astrological underpinning to Horos' arguments: but I offer a more detailed and systematic analysis in this chapter.

encounter a street astrologer. Interpreted in this way, Horos' words at 71–80 mark the crucial opening gambit of an astrological sales pitch: first, having eavesdropped on the conversation between Propertius and the *hospes*, he attracts the poet's attention with a stark warning (71–4); now face to face with a potential client, Horos needs to establish his credentials quickly before his customer walks away (75–80).

Having informed the poet of the apparent truth contained within his books, Horos compares his craft with the unscrupulous behaviour of others before offering a taste of the astrological flavour of his books (79–86):[13]

> di mihi sunt testes non degenerasse propinquos,
> inque meis libris nil prius esse fide
> (nunc pretium fecere deos et fallitur auro
> Iuppiter): obliquae signa iterata rotae
> felicisque Iovis stellas Martisque rapacis
> et grave Saturni sidus in omne caput,
> quid moveant Pisces animosaque signa Leonis,
> lotus et Hesperia quid Capricornus aqua.

> The gods are my witness that I have not disgraced my relatives, and that in my books there is nothing before the truth (now men have made profit of the gods and Jupiter is deceived for gold): the constellations of the slanting circle that constantly revolve, and the lucky star of Jupiter, the rapacious star of Mars, and the star of Saturn severe to all alike, what the Fishes and the lively constellation of Leo cause, as well as Capricorn that is washed in western waters.

Horos refers at random to three planets (83–4) and three signs of the zodiac (85–6—the constellations that revolve on the slanting ecliptic (82)) which are accessible from 'his books' (*meis libris*, 80).[14] This sentiment provides a window onto the competitive arena of the street astrologer, an elusive aspect of the ancient world which has been greatly aided by some illuminating finds among the discarded papyri from Oxyrhynchus.[15] A large number of astrological materials, both

[13] Lines 81–6 are very difficult to interpret, and have typically given rise to transposition of many of the couplets (see esp. Goold 1990) or suggestions of a lacuna (see e.g. Heyworth 2007a). I find most attractive the punctuation proposed by Montanari Caldini (1979) 25–6 and Murgia (1989) 266–7.

[14] I find too far-fetched Keyser's (1992) astrological theory that these lines allude to an accurate horoscope for Propertius' birth date, calculable to May-June 43 BC; for criticism, see esp. Butrica (1993).

[15] See Jones (1999).

horoscopes and astrological charts, have been identified, dating from the 1st century BC onwards, although the majority cluster around the 3rd century AD. The horoscopes range from short, imprecise predictions written crudely on scraps, to more astronomically precise, verbally flamboyant ('deluxe') horoscopes composed on expensive materials. The differing levels of accuracy from these horoscopes point to a range of astrological charts used in their calculation, from the crudity of commercially produced sign-entry almanacs to the more precise primary tables, which would allow a mathematically competent astrologer to furnish a horoscope with specific degrees and minutes.[16]

This opens up, albeit elusively, a world in which astrologers have different levels of skill and/or offer different levels of service dependent on client and financial transaction, all of which might shed further light on what Horos is aiming to do in our passage. By impressing upon Propertius the quality and accuracy of *his* astrological books—books which are equally informative about planets and constellations—Horos seeks to claim a high place for himself within this hierarchy of commercial astrological activity.

Having hoped to dazzle the potential client with his wide-ranging astrological expertise in 81–6, Horos now pursues a different mode of persuasion as he recounts two accurate prophecies that he has previously made (89–102):[17]

> dixi ego, cum geminos produceret Arria natos
> (illa dabat natis arma vetante deo);
> non posse ad patrios sua pila referre Penates.
> nempe meam firmant nunc duo busta fidem.
> quippe Lupercus, equi dum saucia protegit ora,
> heu sibi prolapso non bene cavit equo:
> Gallus at in castris, dum credita signa tuetur,
> concidit ante aquilae rostra cruenta suae.
> fatales pueri, duo funera matris avarae!
> vera, sed invito, contigit ista fides.
> idem ego, cum Cinarae traheret Lucina dolores
> et facerent uteri pondera lenta moram,
> 'Iunoni facito votum impetrabile' dixi.
> illa parit; libris est data palma meis.

[16] See Jones (2007) 309–11. For the variety of apparatus used by professional astrologers in Greco-Roman Egypt, see also Evans (2004).

[17] For the transposition of lines 87–88 to somewhere in Prop. 4.1a, most likely after line 52, see e.g. Murgia (1989).

> I myself said, when Arria was leading out her twin sons (to war)—she was giving her sons arms against the wishes of a god—that they would not be able to bring back their own javelins to their family home: sure enough, two graves now confirm my truthfulness. To be sure Lupercus, while he was protecting the wounded face of his horse, did not take good enough care of himself—alas!—once his horse had fallen forwards. Gallus, on the other hand in the camp, while he was guarding the standards entrusted to him, fell before the blood-stained beak of his own eagle. Ill-fated boys, two funerals caused by a mother's greed! That truthful confirmation came to pass, though I wish it had not. The same person, yours truly, when Lucina was dragging out the labour pains of Cinara and the burden of her womb was slow and creating delay, said, 'Make a vow that is attainable from Juno': the woman gave birth; the palm of victory was granted to my books.

So Horos had predicted the death of Arria's twin sons and the success that would attend Cinara if she prayed to Juno Lucina. He may have been right on these occasions, but his words do little to prove that he is an expert astrologer. The problem is that Horos offers no indication of any astrological basis on which he may have arrived at his predictions:[18] the only supernatural hint he offers is in the case of Arria, where he suggests that a god was forbidding (90 *vetante deo*), a phrase which strikes a religious rather than astrological note.[19] In fact, it might be felt that a prediction that men will die in warfare, and advice to consult Juno during childbirth, are so common in the ancient world as to render Horos a quite unremarkable prophet: any fool might have said these things. Ironically, the proofs of expertise that Horos advances only have the effect of revealing the charlatan behind the guise of competent astrologer.

[18] It might be interesting, however, to consider what would happen in 94 if we replaced *equo* with *Equo*, the constellation of the Horse. Lines 93–4 would then constitute an ironic comment from the astrologer on the danger of putting mortal before astrological concerns: 'To be sure Lupercus, while he was protecting the wounded face of his horse, did not take good enough care of himself when the Horse had set (lit. fallen forwards, into the sea)'. Lines 95–6 also contain terminology which is equally appropriate in the astrological sphere (*signa*, *aquila/Aquila*), but the surrounding context makes an astrological reading far less likely. For differing speculative attempts to construct an astrological reading of these lines, see Dieterich (1900) 212–13; Montanari Caldini (1979) 32–6.

[19] The phrase may be used deliberately to create a connection with Propertius, who is likewise advancing against the wishes of a god, Apollo (73): Propertius, then, should take special notice of this example of the dire consequences of proceeding without divine sanction.

Horos' subsequent celebration of astrology as the most important form of divination (103–8)—surpassing the Libyan oracle of Jupiter (103), haruspicy (104), augury (105) and magic, in the form of hydromancy (106)—might certainly provide us with contemporary evidence of the increasingly important role that astrology was occupying in imperial Rome as measured against more traditional forms of divination.[20] But coming as it does after two ineffective examples in 89–102, it does little to restore our confidence in the astrological competence of the speaker.[21]

After a digression on Calchas' famous decision to sanction the sacrifice of Iphigeneia (109–18)—a story perhaps designed to advise the poet against risking long-term disaster for short-term gain—Horos turns at last to discussion of Propertius' own life (119–46):

hactenus historiae; nunc ad tua devehor astra.
    incipe tu lacrimis aequus adesse novis.
Vmbria te notis antiqua Penatibus edit
    —mentior? an patriae tangitur ora tuae?
qua nebulosa cavo rorat Mevania campo
    et lacus aestivis intepet Vmber aquis?—
[scandentisque Asis consurgit vertice murus,
    murus ab ingenio notior ille tuo.][22]
ossaque legisti non illa aetate legenda
    patris, et in tenues cogeris ipse lares.
nam tua cum multi versarent rura iuvenci,
    abstulit excultas pertica tristis opes.
mox ubi bulla rudi dimissa est aurea collo,
    matris et ante deos libera sumpta toga,
tum tibi pauca suo de carmine dictat Apollo,
    et vetat insano verba tonare foro.
'at tu finge elegos, fallax opus—haec tua castra—
    scribat ut exemplo cetera turba tuo.
militiam Veneris blandis patiere sub armis,
    et Veneris pueris utilis hostis eris.
nam tibi victrices quascumque labore pararis,
    eludet palmas una puella tuas.

[20] See Chapter 3.

[21] Horos does not help himself here by making a serious error in 105 *aut si quis motas cornicis senserit alas* ('or if anyone should have observed the movement of the crow's wings'): in augural terms the crow is one of the *oscines*, a bird which offers predictions through its cries, not its movement; cf. Fest. 214 Lindsay.

[22] Lines 125–6 are deleted from Heyworth (2007a) and are deemed spurious by Hutchinson (2006). For my reasons for reinstating them, see discussion later.

et bene cum fixum mento decusseris uncum,
nil erit hoc: rostro te premet ansa suo.
illius arbitrio noctem lucemque videbis;
gutta quoque ex oculis non nisi iussa cadet.
nec mille excubiae nec te signata iuvabunt
limina: persuasae fallere rima sat est.'

So much for history: now I am borne towards your stars. Prepare to attend with composure to your new tears. Ancient Umbria brought you forth in a renowned household—am I telling a lie? Or have I hit upon the border of your native land, the place where misty Mevania sheds its dew on the sunken plain and the Umbrian lake is warm with its summer waters [and the wall of soaring Assisi rises up on the hilltop, that wall more famous because of your talent]? You gathered the bones of your father, at an age when they ought not to have been gathered, and you yourself were forced into a humble home. For at a time when many a bullock was ploughing your fields, the severe measuring-rod of the surveyor stole away the property you had cultivated. As soon as the golden locket had been removed from your innocent neck and you had assumed the toga of manhood before your mother's gods, at that time Apollo dictated to you a few words from his own prophecy and forbade you to thunder down words in the mad forum. 'Instead, go and compose elegy, a deceptive task—for this is your camp—so that the rest of the crowd may write following your example. You will suffer warfare under the coaxing arms of Venus, and for Venus' boys you will be a profitable adversary. For whatever symbols of victory you prepare for yourself by your toil, a single girl will mock your triumphs. And when you dislodge the well-fixed hook from your chin, it will be of no use: the handle will hold you with its own curvature. Subject to her direction will you see both night and day; not even a teardrop will fall from your eyes unless it has been bidden. Nor shall a thousand watches help you, nor a thousand seals on her threshold: a chink in the door is enough for a woman resolved to deceive.'

Horos here presents Propertius with a summary of the content of his own horoscope in chronological order: his upbringing (121–6), misfortunes of family and estate (127–30) and his choice of poetic career on becoming a man (131–4), influenced apparently by words from Apollo telling him that he should not/cannot stray from the love of a woman and from the realm of love elegy (135–46), a lifestyle clearly identifiable by allusions to *servitium amoris* (139–44) and the *paraclausithyron* (145–6).

Horos' narration is clearly designed to further win the confidence of his potential client by mentioning aspects of his life that the astrologer could not have previously known. Does Horos achieve this objective? On the positive side, it would appear that the astrologer is indeed correct about Propertius' origins in Umbria—confirmed by the author himself at 1.22.9–10—and mention of a loss of patrimony (4.1B.129–30) might elaborate on earlier comments from the poet about his paltry estate.[23]

By now, however, we should be wary of this potential confidence trickster.[24] Tellingly, as with the cases of Arria and Cinara (89–102), Horos again offers no clue as to the astrological basis on which he has acquired his information about Propertius' life. Horos' presentation may have the general flavour of a horoscope, to judge by the examples preserved in Vettius Valens: change in financial circumstances,[25] oscillation between a position of power and slavery,[26] the death of a family member,[27] and even (rapid) reversals in fortune specifically caused by a woman,[28] are topics attested in his horoscopes. But such topics in Vettius' horoscopes are (dryly) underpinned by precise information about positions of constellations and planets. The absence of any astrological data from Horos' horoscope is, therefore, even more striking, and invites the reader to entertain other, less spectacular explanations to account for Horos' acquisition of knowledge. I argued earlier that Horos' words in 71–4 betray the fact that he has overheard the final words of Propertius in 1A.67–70. It seems a small step to suggest that he also overheard the poet's words immediately prior to this (1A.61–66):

> Ennius hirsuta cingat sua dicta corona:
> mi folia ex hedera porrige, Bacche, tua,
> ut nostris tumefacta superbiat Vmbria libris,

[23] Cf. 2.34.55 *aspice me, cui parva domi fortuna relicta est* ('look at me, to whom was left a scanty fortune at home'); also 2.24.35–8 (imagined funeral speech) *'ossa, Properti,/ haec tua sunt . . . quamvis non ita dives eras'* ('these are your bones, Propertius, . . . even though you were not so rich').

[24] As will become apparent from the forthcoming discussion, I believe it to be potentially perilous to disregard the difference in primary narrator in Prop. 1.22 and 4.1B.119–46 and simply use Horos' comments on Propertius' life to flesh out the poet's own autobiographical details elsewhere in the corpus.

[25] Cf. e.g. Vet. Val. 2.21 with Neugebauer and Van Hoesen (1959) 97.

[26] Cf. e.g. Vet. Val. 2.21 with Neugebauer and Van Hoesen (1959) 86.

[27] For the death of a child, cf. e.g. Vet. Val. 5.10 with Neugebauer and Van Hoesen (1959) 103.

[28] Cf. Vet. Val. 7.5 with Neugebauer and Van Hoesen (1959) 107; Vet. Val. 5.10 with Neugebauer and Van Hoesen (1959) 111.

Vmbria Romani patria Callimachi!
scandentes quisquis cernet de vallibus arces,
ingenio muros aestimet ille meo.

Let Ennius surround his verses with a shaggy crown: O Bacchus, stretch out for me leaves from your ivy so that Umbria may swell with pride at my books, Umbria, the fatherland of the Roman Callimachus! Whoever shall look upon the citadels which climb up from the valleys, let him esteem the walls by my genius.

Here, Propertius hopes that his projected aetiological poem will make his homeland Umbria proud, and that its walls will be made famous by his genius.[29] As this is the only information that Propertius has given about himself in 1A, it is not surprising that Horos should mention it first as a way of reeling in his would-be client.[30] Line 122 is telling, as Horos directly asks Propertius if he thinks he is lying (*mentior*?): it is particularly apt that the wily Horos should elect the one piece of information he knows to be true to be the index of his authority. That Horos is little more than a clever eavesdropper is particularly apparent in 125–6, where the speaker flatteringly quotes back to Propertius the sentiment for which the poet himself had hoped in 65–6: note the repetition of *murus*, *ingenio* and the present participle of *scando*.[31] But there is a twist: Propertius hopes that his aetiological project will bring fame to Umbria; Horos suggests that he has *already* achieved this, namely through the medium of love elegy, a sentiment which will form part of his warning to the poet not to stray from this genre.

Horos, then, has gleaned the information about Propertius' Umbrian origins not from astrology—which would seem to have been impossible in any event, given that one star was believed to govern all of Italy (cf. Man. 4.773–7)—but by adopting more underhand tactics.[32] What about the rest of the information he presents? Most of the

[29] I follow the prevailing view that Propertius must be referring to valleys and citadels in Umbria in 65–6; note that he switches his attention to Rome in line 67.

[30] I take lines 123–4 to be a general expansion of the geography of Umbria ('Umbria, the place where . . .') rather than a specific pinpointing of Propertius' birthplace by Horos.

[31] Repetition between 65–6 and 125–6 is an important part of the make-up of Horos' characterisation. For different readings of the force of the repetition, see Stahl (1985) 273–4; Wills (1996) 149–51.

[32] A charlatan astrologer's underhand capacity for acquiring some knowledge of their client is something about which the later philosopher Favorinus warns his Roman audience (Gel. 14.1.33). See further Chapter 11.

claims Horos makes—on the death of Propertius' father (127–8) and the dictates apparently given to him by Apollo to pursue a career as an elegiac poet rather than a politician/lawyer (131–46)—cannot be substantiated either way. Propertius makes no mention of these details elsewhere in his corpus—they are notably absent from his only autobiographical poem, 1.22—and we can only guess at his likely reply to Horos' speech (the poem finishes as soon as Horos stops speaking). Does the fact that Propertius nowhere concords with Horos' remarks show the astrologer to be way off the mark? Finally, it is worth looking more closely at Horos' comment about the slender patrimony of the poet (129–30). It was said earlier that it might be read as an expansion of what Propertius himself admits at 2.34.55. But we should not be blind to the difference between the two passages: in Book 2, Propertius simply says that he was left a small fortune, and does not mention, as Horos does, that this resulted from a confiscation of land. Perhaps Horos is speaking the truth here—we cannot tell, as Propertius nowhere deals with this aspect of his life—but he may well be simply using external knowledge to create the illusion of divinatory powers: any intelligent individual would have known that Octavian's warfare and, consequently, land confiscations were centred around the city of Perusia in Umbria in 41–40 BC.[33] The suggestion that Propertius was himself affected by this might, then, constitute a calculated gamble on the part of the charlatan.

Having now dealt with the poet's past, Horos finally turns to the poet's future, with a specific warning which, at last, appears to have an astrological thrust (147–50):

> nunc tua vel mediis puppis luctetur in undis,
> vel licet armatis hostis inermis eas,
> vel tremefacta cavos tellus diducat hiatus,
> octipedis Cancri terga sinistra time.

> Though your ship may now be struggling in the middle of the ocean, though you may be proceeding as an unarmed enemy against armed men, though the earth may tremble and open up deep chasms, fear the unfavourable back of the eight-footed Crab.

The significance of this final, astrological warning from Horos has exercised scholars over the years, and no consensus is possible. As

[33] Numerous sources make reference to this; cf. e.g. Suet. *Aug.* 13–15; App. BC 5.48–9; Dio 48.13–15; see further Hutchinson (2006) 83.

Cancer is one of the two turning points in the sun's yearly journey, it may be particularly apt as part of a warning about changing (poetic) direction. The sign might hold more specific potency for the elegiac lover/poet, as Cancer is associated with lust and avarice (key traits of the *puella*) and separation (the key fear of the lover).[34] Others have argued that the thrust of Horos' final warning may not be astrological at all. In the light of the fact that some contemporary (*c.* 16 BC) gold and silver coinage displayed a Crab on the reverse, Horos may be warning the poet against pursuing short-term financial gain at the expense of long-term tragedy.[35] Alternatively, this may be a literary allusion to Gallus' submission to love elegy, a poet who had recognised that he could not change poetic course (to pastoral poetry) under the star of Cancer (Verg. *Ecl.* 10.68).[36] In summary, it is possible to take Horos' final words as a serious, multi-layered warning. In the context of his extravagant boasts and detailed attempts to establish credentials, however, I am inclined to read this as a final debunking of Horos: despite his bravado, the charlatan astrologer can only muster one extremely nebulous piece of advice for the future, which relies on the credulity of the listener to interpret for himself.

The figure of Horos is undoubtedly tied in to Propertius' literary ambitions and reservations concerning his fourth book of poetry. It is equally valid to say, however, that Horos is developed as a character in his own right, as a foreign astrologer in the streets of Rome, and should not be simply reduced to either a novel mouthpiece for the god of poetry, Apollo, or an alter-ego for the cautious poet. As the most detailed depiction of the street astrologer to have come down to us from antiquity, it is all the more important that we do not neglect historical and socio-political readings of this Augustan text. In general terms, the fact that an astrologer was chosen for the divinatory role in this Rome-centred poem, and that astrology is specifically afforded a vaunted position among the various forms of divination (103–8), points to the growing influence of and respect for astrology in contemporary imperial Rome. But Propertius' respect for the art itself

[34] For various astrological readings of this warning, see Nethercut (1968), (1970); Marquis (1973); Montanari Caldini (1979) 73–85; see also Hutchinson (2006) 85–6.

[35] See Stahl (1985) 278–9.

[36] See Kaufhold (1996).

does not go hand in hand with respect for all its practitioners.[37] On the contrary, Propertius has chosen to depict in amusing fashion one of its less scrupulous exponents, an eavesdropping confidence trickster and salesman who accosts the poet in the street[38] and who, despite his astrological bravado, manages to offer no tangible evidence of his ability to predict the truth from reading the stars.

[37] I disagree with Courtney (1969: 73) that Prop. 2.27.1–4 shows disrespect for astrology: the context of the poem makes it clear that Propertius is criticising human anxiety about the future rather than any form of divination.

[38] An important point: as it is Horos who first engages the poet in conversation, the poet is absolved from any politically awkward accusation of having actively consulted an astrologer about the future. Propertius' poem, then, provides an astrological equivalent to Horace's *Satire* 1.9, in which the poet is similarly accosted in the streets of Rome by a novice eager to prove his credentials.

11

Conclusion

Manilian Dialogues and the Relaxing of Astrological Discretion

In this study, it has been my aim to explore the varied discourses in disclosure and discretion that authors under Octavian/Augustus adopt in their treatment of the politically sensitive topic of astrology. Now that all these authors have been covered in the preceding chapters, I am in a position to offer, in the summary that follows, a chronological approach to emerging literary strategies in this regard.

While Cicero appears to bear witness to the changing status for astrology in the wake of Caesar's comet, it is arguably Vitruvius who confronts more directly the issue of the parameters of appropriate stellar discourse in an early Augustan context which has already witnessed the Emperor's mixture of personal astrological celebration (in the form of an early circulation of his horoscope) with restrictions on its use by others (most notably in the form of the expulsion of astrologers in 33 BC). Vitruvius' choices, simply put, are to acknowledge Caesar's astral deification, to afford exposure to the stellar backdrop for a study of astrology through a close engagement with Aratus, and to acknowledge the causal power of stars over both the weather and the lives of mortals, while strictly deferring the practice of astrology to experts. Many of these early strategic choices from Vitruvius are taken up and developed by later authors. We have witnessed the dynamic mythical development of Caesar's astral deification, the continuing popularity of Aratus and the creativity with which he is deployed, and the emergence of active avoidance of detailed astrological disclosure as a literary manifesto. Vitruvius is, however, implicitly acknowledged as over-cautious by some, who separate out astrometeorology from natal astrology and find in the former an appropriate means of advertising the causal force of the stars without straying into inappropriate imperial territory. Others still, like Propertius, take a different approach by drawing attention to the potential discrepancy between a respectful science and the fraudulent nature of some of its practitioners.

To return finally and briefly to Manilius, the heart of this study and most elaborate exponent of disclosure and discretion, one can, on the strength of the analyses in this book, posit a number of potential dialogues between Manilius and his predecessors, of which the following two brief case studies claim only to provide a flavour.

MANILIUS, OVID AND GERMANICUS: CONTEMPORARY STRATEGIES FOR ROMANISING ARATEAN STARS

Although the relative dates for Manilius, Ovid (*Fasti*) and Germanicus are destined to remain a matter of scholarly debate, one can at least agree that they are all writing against the backdrop of a late Augustan, and possibly early Tiberian, political environment.[1] As such, one can talk with more conviction in terms of the shared intellectual space in which these Roman authors are operating. Moreover, if we look a little more closely, we can detect an air of competition and rivalry as each author seeks to find an appropriate Roman means of celebrating the power of the stars.

Aratus is key to all three texts. Manilius and Germanicus in particular acknowledge the importance of Aratus by making extensive and sustained use of *Phaenomena* for (at least) the starting point to their works (Germ. 1–725, Man. 1). More sporadically, for Ovid, Aratus looms large in the observations on risings and settings of constellations which form an important part of the poet's religious project in *Fasti* (1.2, 295–6).[2] But for all three, Aratus is modified in different ways to suit both personal stylistic and Roman political requirements. To their Aratean frame, both Germanicus and Ovid choose to supplement a significant number of catasterisms and revel in mythical detail, sometimes of a very erudite nature.[3] By contrast, Manilius'

[1] For the relative dating of Manilius and Germanicus, see Abry (1993) and Colborn (2013). For the dating of Ovid's *Fasti*, see Green (2004) 15–25.

[2] See e.g. Green (2004) 144–7.

[3] See Green (2004) 145–7; Possanza (2004) 170–217.

professed disdain for catasterism narratives (2.25–38, despite the Andromeda epyllion in Book 5) betrays a general tendency to shy away from mythological ornament to his instruction.[4] But Aratus is also reconfigured to serve a variety of late Augustan political purposes. In particular, for Ovid, Aratus is pressed into the service of celebrating the Emperor's newly restored Julian calendar (*c.* AD 4–8). Not only is Aratus 'fragmented' to fit the temporal demands of the new calendar;[5] Ovid ventures further on occasions by rejecting altogether the Aratean system of constellations as signals for weather in favour of a system whereby both constellations and weather act as heralds for the new Roman date.[6] More interesting for our purposes are the divergent ways in which the politicisation of astrology is acknowledged in each poem. Germanicus, as we have seen, stretches the Aratean field of astrometeorology to its limit by assigning a causal force to zodiacal and planetary signs over weather patterns on earth, while avoiding the branch of genethliological astrology which impinges on the daily lives of mortals. Manilius, by contrast, adopts a quite traditional stance towards astrometeorology—only the Dog Star is afforded a causal role over the weather (1.396–411)—and supplants Aratus' astrometeorological material with a celebratory exploration of the power of the stars over human beings, albeit in a manner which denies his readers the full knowledge to become practitioners. One particular tactic, an unfulfilled pledge to discuss the planets, binds both Manilius and Ovid.

MANILIUS AND CICERO: THE DEVELOPMENT OF PHILOSOPHICAL ARGUMENTATION ON ASTROLOGY

If the political will towards horoscopic astrology changed radically between the times of Cicero and Manilius, there was a corresponding

[4] See Abry (2007a) 6–13.

[5] Ovid's astronomical methodology for assigning dates to the rising and setting of constellations, a topic traditionally ridiculed by scholars following in the footsteps of Ideler (1825), has recently been the subject of more charitable assessments; see Fox (2004).

[6] Cf. e.g. 1.315–16 *institerint Nonae, missi tibi nubibus atris/signa dabunt imbres exoriente Lyra* ('if the Nones are at hand, showers sent from black clouds will provide signs for you, as the Lyre is rising'), possibly also 4.901–4.

need in philosophical circles to respond to popular arguments made against the science. Manilius' project does, to some extent, respond to and seek to head off the philosophical objections levelled against astrology by his predecessor in *De Divinatione*.

At *Div*. 2.92–3, Cicero criticises astrologers for not taking into account the role of geographical location in a reading of the stars: as different parts of the world bear witness to a different set of rising stars, it is ignorant to maintain that one set of conditions governs all those born on a particular day, irrespective of location. While this form of argument remained popular with later opponents of astrology, Manilius addresses the concern by drawing attention to the peculiar factors—both climatic and stellar—which grant particular characteristics to entire nations (4.711–817), quite separate from the horoscopes of individuals.[7] While Cicero objects to astrology on the basis that individuals born at the same moment (and presumably in the same place) experience different lives (*Div*. 2.95), Manilius counters by proffering a more nuanced astrological system whereby the twelve divisions within a single sign (dodecatemories) exert a more complex set of influences, resulting in different lives for those born under the same sign (2.693–712, esp. 707–12).[8]

The change in emphasis from Cicero to Manilius is, however, most acutely felt in the use to which each author puts the horoscope of Rome, famously cast by L. Tarutius Firmanus, who calculated that Rome was born under the guardianship of Libra. In Cicero (*Div*. 2.98–9), Tarutius is a figure worthy of light ridicule who is suitably distanced from proper Roman conduct.[9] The same horoscope from Tarutius, however, receives the utmost respect in Manilius, now that Libra carries additional imperial weight by virtue of its association with an Emperor (4.769–77).[10] How times have changed, as Cicero's ultimate example of astrological foolishness provides the impetus in Manilius for a high form of imperial flattery.

[7] For later exponents of this argument against astrology, cf. Gel. 14.1.7–11 (Favorinus), Sext. Emp. *Math*. 5.83–5. Long (1982: 175–7) suggests that Manilius is the first to develop properly the case of ethnographic astrology specifically to counter such objections.

[8] Once again, Manilius stands out here as a voice against continuing scepticism in this regard; cf. Sext. Emp. *Math*. 5.88–9, 99.

[9] See Chapter 4.

[10] For the vexed question of the identity of the Emperor, see Volk (2009) 156–9.

ASTROLOGICAL RUMBLINGS: THE RELAXING OF DISCRETION AFTER AUGUSTUS

After an era of respect and restraint under Rome's first Emperor, a less charitable assessment of both the astrologer and his science re-emerges in written form with Seneca the Elder. Composed over a short space of time in the 30's AD, possibly after Caligula's accession in AD 37, Seneca's declamatory works (*Controversiae* and *Suasoriae*) make frequent allusion to events under Tiberius.[11] And for a man with the potential to look unfavourably upon the early empire, it is not difficult to understand how the increasing role of astrology in the exercising of (abusive) imperial power might have provided Seneca with a fitting focus of attack.[12]

Suasoriae 4 takes the form of a declamation from Arellius Fuscus, teacher of rhetoric for Ovid and heralded by Seneca as one of the four best declaimers of his age (*Con.* 10 *praef.* 13). For this particular exercise, Arellius takes on the role of advisor to Alexander the Great, providing guidance as to whether or not he should enter Babylon following a warning from an augur. The exposition takes the form of a mocking attack on divination itself, focussed around not the augur but the figure of the astrologer: both the astrologer and astrology become subjects of ridicule, as Arellius ultimately brands the 'science' nothing more than a confidence trick (*Suas.* 4.3):[13]

> incertae enim sortis vivimus. unicuique ista pro ingenio finguntur, non ex <scientiae> fide.
>
> For uncertain is the lot by which we live. Those things [sc. astrological predictions] are conjured up according to the disposition of the particular individual, not out of trust in a science.

Earlier in the declamation, Arellius finds various means of mocking the 'supernatural' status of the astrologer (*Suas.* 4.1):

> magnus iste et supra humanae sortis habitum sit cui liceat **terrere** Alexandrum; **ponat iste suos inter sidera patres et originem caelo trahat**,

[11] See in general Griffin (1972) 11; Sussman (1978) 23–4; Fairweather (1981) 15.

[12] Seneca's general attitude towards contemporary society is one of moral decline (*Con.* 1 *praef.* 8–10); he laments contemporary limits to literary free speech (*Con.* 10 *praef.* 5–7), in contrast apparently to the habit of Augustus (*Con.* 2.4.13); and several of his friends were persecuted by the Emperors. See further Griffin (1972) 4–17; Sussman (1978) 28–33; Fairweather (1981) 8–14.

[13] The text of Seneca is taken from Håkanson (1989).

agnoscat suum vatem deus. non eodem vitae fine aetatem agat; extra omnem fatorum necessitatem caput sit, quod gentibus futura praecipiat.

Let him be great indeed and beyond the disposition of human lot, that man who has the power to **terrify** Alexander. **Let him place his own ancestors/fathers among the stars and draw his own origin from heaven**, let a god recognise his own prophet. Let him live his days with a different end of life [to the rest of us]; and outside all the necessities of the fates, let him be the font that can teach the future to the nations.

The entire episode is, of course, a complex of times, as Arellius (born *c.* 60–55 BC) advises the great Greek general before his fateful incursion into Babylon (323 BC), and the complexity increases when we examine the specific elements in the declamation. Irrespective of when Arellius' declamation was supposed to have been delivered orally, Seneca's written version is given a conspicuously imperial, not to say Tiberian, colouring. The extant sources for Alexander offer no detail as to the precise nature of the warning given by the astrologers, focusing instead on Alexander's initial disregard for them and suspicion of their mercenary motives.[14] The specific elements of Arellius' attack appear to arise, rather, from a contemporary imperial context. Most obviously, mockery of an individual who ventures both to claim divine descent and to place his fathers among the stars can only reflect negatively on the actions and pretensions of both Augustus and Tiberius in their dealings with their 'fathers'.[15] More subtly, the witty reversal of hierarchies, whereby Alexander loses his traditional epithet to the astrologer (*magnus*) to become his fearful subject (*terrere*), creates a scenario which recalls in particular the anecdotal evidence for Tiberius' nervous subservience to Thrasyllus during his reign.[16]

[14] Cf. Plu. *Alex.* 73.1–5; Arr. *Anab.* 7.16.5–17.6.

[15] For the motif of Octavian raising Caesar to the stars, see Chapter 9. The Senate voted divine honours to Augustus shortly after his death in AD 14 (Tac. *Ann.* 1.10), a move clearly endorsed by Tiberius, who refers to his father as 'divus Augustus' (Tac. *Ann.* 1.11).

[16] Though fear is conceded as a possible reason for Augustus' reticence in the presence of the astrologer Theogenes (Suet. *Aug.* 94.12 *metu ac pudore*), fearful deference to an astrologer fits more readily the profile of Tiberius, who consulted the wisdom of Thrasyllus for a range of political matters, including issues of succession (Suet. *Cal.* 19.3), potential rivals (Tac. *Ann.* 6.20) and life expectancy in the face of omens (Dio 58.27.1–3); it is at least possible that Thrasyllus lay behind Tiberius' decision to retire from Rome in AD 26; see further Chapter 5 and Cramer (1954) 104–8. By contrast, Alexander is never 'terrified' by astrologers, although he may later express regret for his initial mistrust of them (Plu. *Alex.* 73.1–5).

In Neronian times, the special status afforded to Caesar's comet, safe under Augustus, comes under attack from a contemporary writer.[17] Calpurnius Siculus offers, through the figure of Faunus, a positive interpretation of a comet of AD 54 as a sign of Nero's accession, and in the process offers a comparison with its destructive Caesarian counterpart (*Ecl.* 1.77–83):[18]

cernitis ut puro nox iam vicesima caelo
fulgeat et placida radiantem luce cometem
proferat? ut liquidum niteat sine vulnere plenus?
numquid utrumque polum, sicut solet, igne cruento
spargit et ardenti scintillat sanguine lampas?
at quondam non talis erat, cum Caesare rapto
indixit miseris fatalia civibus arma.

Do you see how already for a twentieth time the night gleams in a cloudless sky, and brings forth a comet beaming with tranquil light? Do you see how clearly it shines, burning brightly with no presage of bloodshed? Surely it is not with any bloody fire that the torch sprinkles each pole, as is the way of comets, and it is not with any burning blood that it flashes? But once upon a time it was not like this, when, with Julius Caesar snatched away, it pronounced fatal wars to wretched citizens.

In this way does Calpurnius herald a new Emperor and a new dawn, while Caesar's comet, once regarded as special under the reign of his son Augustus, falls back in line with its traditional counterparts that presage major civic upheaval.[19]

Our final stop on this brief journey into the afterlife of Augustan astrological discretion centres around the mercurial figure of Favorinus (*c.* AD 85–165).[20] An Academic philosopher from Gaul and a Roman *eques* who spent most of his life at Rome, Favorinus emerged as one of the leading intellectual figures under Hadrian. In close circles with both Plutarch and Dio, he was especially praised and respected by Aulus Gellius, who purports to recall, sometimes verbatim, many of his declamations in *Attic Nights*.[21]

[17] The following discussion assumes Calpurnius Siculus to be an early Neronian poet; for the continuing debate, see esp. Horsfall (1997).

[18] The text of Calpurnius Siculus is taken from Duff and Duff (1934).

[19] See Chapter 3.

[20] For varying attempts to piece together Favorinus' career and intellectual circle, see Swain (1989); Gleason (1995) 3–20; Bowie (1997); Holford-Strevens (1997).

[21] Moreover, as a multidisciplinary figure, Favorinus fits the intellectual program of *Attic Nights* and may have provided a model for its brand of practical polymathy. See Beall (2001).

Favorinus' declamation against astrologers, recounted in *NA* 14.1, is apparently based on memory and accompanying written notes from a live performance at Rome that Gellius had attended (14.1.2), although the author's access to a written copy may also be implied by the extensive tracts of direct speech quoted during the section.

The speech bears the hallmarks of Favorinus' Academic stance in its preference for different forms of persuasive argumentation over consistency of standpoint. The principles of astrology itself come under attack, for example, when the speaker advances hereditary factors behind a human's make-up (14.1.21–2) and asks why animals are not included in astrological predictions (14.1.31), and why people born at different times suffer the same fate while no two people are ever identical (14.1.27–9); behind all this lie the false machinations and deceptions of the charlatan practitioner (14.1.2–3, 32–3). At other times in the same speech, however, Favorinus appears to accept the validity of the science but to question the ability of any mortal to be able to grasp such a complex and elusive art. No one, says the speaker, has the intellect to master astrology (14.1.5), to take account of geographical variations in stellar positioning (14.1.8–10), to see all the relevant stars (14.1.11–13, 26) and to choose the most appropriate moment from which to draw a horoscope (14.1.19–21). So diverse is the argumentation that Gellius himself cannot be sure whether Favorinus speaks from conviction or a desire for intellectual display (14.1.2):[22]

> exercendine aut ostentandi gratia ingenii, an quod ita serio iudicatoque existimaret, non habeo dicere.

> But whether it was for the purpose of exercising or flaunting his talent, or because he thought seriously and decisively along these lines, I am unable to say.

At a formal level, one can view Favorinus' declamation as a conglomeration of many of the arguments and strategies that have gone before—providing as such a fitting end to this study. Favorinus mixes (pre-Augustan) Ciceronian philosophical argumentation against the science itself with more implicit (Augustan) Propertian and Manilian concerns about the inadequacies of the would-be astrologer.[23]

[22] The text of Aulus Gellius is taken from Marshall (1990).

[23] For Ciceronian-style argument in Favorinus, cf. Gel. 14.1.8–10 with Cic. *Div.* 2.92–3 (geographical variation); Gel. 14.1.26–9 with Cic. *Div.* 2.90, 95, 97 (lives different though born at the same time, and common deaths for those born at different times); Gel. 14.1.31 with Cic. *Div.* 2.98 (astrology does not extend into the animal kingdom). See also Long (1982) 184–5.

If it is not possible to date Favorinus' live performance (it is often assumed to be Hadrianic),[24] it is equally impossible to recover its audience's reaction. Hadrian, to be sure, is depicted as an active enthusiast of astrology during his reign, at least in comparison to his predecessor Trajan. Moreover, Favorinus famously advertises himself as a man with the power to flout imperial sensibilities: he quarrelled with an Emperor and lived (Philostratus, *VS* 489). Did attacks of this sort on imperially sensitive topics cause serious animosity between the two, perhaps paving the way to Favorinus' exile, or did their relationship rarely rise above the level of playful intellectual banter? Scholars lean towards the latter interpretation.[25] With restraint cast aside, high-profile criticism of astrology, delivered in a Roman intellectual context, had returned once more.

[24] See e.g. Cramer (1954) 197–8.

[25] There is now general scholarly agreement that Favorinus' quarrel with the Emperor was neither long-lasting nor a cause of exile: the work *On Exile*, attributed to Favorinus, is not generally felt to be autobiographical; see Swain (1989); Gleason (1995) 145–58; Bowie (1997) 4–11; Holford-Strevens (1997) 191–9. Dealing specifically with Gel. 14.1, Cramer (1954: 197–8) likewise reads potential interaction between Hadrian and Favorinus on the topic of astrology as quite playful.

Bibliography

Abry, J.-H. (1988) 'Auguste: La Balance et le Capricorne', *REL* 66: 103–21.

——— (1993) 'Manilius et Germanicus: une énigme historique et littéraire', *REL* 71: 179–202.

——— (2007a) 'Manilius and Aratus: Two Stoic Poets on Stars', *Leeds International Classical Studies* 6.1: 1–18.

——— (2007b) 'Les vers 5,710–745 sont-ils l'épilogue des *Astronomiques?*', in B. Bureau and C. Nicolas (eds), *Commencer et Finir dans les Littératures Antiques* (Lyons: Collection du CRGR), 293–333.

——— (2011) 'Cosmos and Imperium: Politicized Digressions in Manilius' *Astronomica*', in S.J. Green and K. Volk (eds), *Forgotten Stars: Rediscovering Manilius Astronomica* (Oxford: Oxford University Press), 222–34.

Allen, J. (2001) *Inference from Signs: Ancient Debates About the Nature of Evidence* (Oxford: Oxford University Press).

Alton, E.H., Wormell, D.E.W. and Courtney, E. (1988) *Ovidius Fasti* (Stuttgart: Teubner).

Anderson, J.C. (1997) *Roman Architecture and Society* (Baltimore and London: Johns Hopkins University Press).

Anderson, W.S. (1977), *P. Ovidii Nasonis Metamorphoses* (Leipzig: Teubner).

Aujac, G. (1975) *Géminos, introduction aux phénomènes* (Paris: Les Belles Lettres).

Austin, R.G. (1977) *P. Vergili Maronis Aeneidos, Liber Sextus* (Oxford: Clarendon Press).

Ax, W. (1977) *M. Tullius Cicero: De Divinatione, De Fato/Timaeus* (Stuttgart: Teubner).

Baier, T. (1999) 'Myth and Politics in Varro's Historical Writings', *EMC* 43 n.s.18: 351–67.

Bailey, C. (1947) *Lucreti De Rerum Natura Libri Sex, Volume I: Prolegomena, Text and Critical Apparatus, Translation* (Oxford: Clarendon Press).

Bakhouche, B. (2002) *L'Astrologie à Rome* (Louvain-Paris: Peeters).

Baldwin, B. (1990) 'The Date, Identity and Career of Vitruvius', *Latomus* 49: 425–34.

Barrett, A.A. (1978) 'Observations of Comets in Greek and Roman Sources before A.D. 410', *Journal of the Roman Astronomical Society of Canada* 72: 81–106.

Barton, T. (1994a) *Power and Knowledge: Astrology, Physiognomics and Medicine Under the Roman Empire* (Ann Arbor: University of Michigan Press).

——— (1994b) *Ancient Astrology* (London and New York: Routledge).

Barton, T. (1994c) 'Astrology and the State in Imperial Rome', in N. Thomas and C. Humphrey (eds), *Shamanism, History and the State* (Ann Arbor: University of Michigan Press), 146–63.

——— (1995) 'Augustus and Capricorn: Astrological Polyvalency and Imperial Rhetoric', *JRS* 85: 33–51.

Beall, S.M. (2001) '*Homo Fandi Dulcissimus*: The Role of Favorinus in the *Attic Nights* of Aulus Gellius', *AJP* 122: 87–106.

Beard, M. (1986) 'Cicero and Divination: The Formation of a Latin Discourse', *JRS* 76: 33–46.

——— and Henderson, J. (1998) 'The Emperor's New Body: Apotheosis from Rome', in M. Wyke (ed.), *Parchments of Gender: Deciphering the Bodies of Antiquity* (Oxford: Oxford University Press), 191–219.

Beard, M., North, J. and Price, S. (1998) *Religions of Rome Volume I: A History* (Cambridge: Cambridge University Press).

Bing, P. (1993) 'Aratus and his Audiences', *MD* 31: 99–109.

Boissevain, U.P. (1955–69), *Cassii Dionis Historiarum romanarum quae supersunt* (Berlin: Weidmann).

Boll, F. (1903) *Sphaera: Neue Grieschische Texte und Untersuchungen zur Geschichte der Sternbilder* (Leipzig: Teubner).

Bollók, J. (1993) 'Horace and the Astrology', *ACD* 29: 11–19.

Bosworth, B. (1999) 'Augustus, the *Res Gestae* and Hellenistic Theories of Apotheosis', *JRS* 89: 1–18.

Bouché-Leclercq, A. (1899) *L'Astrologie grecque* (Paris: Ernest Leroux).

Bowie, E. (1997) 'Hadrian, Favorinus, and Plutarch', in J. Mossman (ed.), *Plutarch and His Intellectual World* (London: Duckworth), 1–15.

Bowman, A.K. (1991) 'Literacy in the Roman Empire: Mass and Mode', in J.H. Humphrey (ed.), *Literacy in the Roman World* (*JRA* supplement 3), 119–31.

Bradshaw, A. (2002) 'Horace's Birthday and Deathday', in T. Woodman and D. Feeney (eds), *Traditions and Contexts in the Poetry of Horace* (Cambridge: Cambridge University Press), 1–16.

Bram, J.R. (1975) *Ancient Astrology, Theory and Practice. The Mathesis of Firmicus Maternus* (ParkRidge, NJ: Noyes).

Brind'Amour, P. (1983) 'Manilius and the Computation of the Ascendant', *CP* 78: 144–8.

Briscoe, J. (1998) *Valeri Maximi Facta et Dicta Memorabilia* (Stuttgart: Teubner).

Brosch, N. (2008) *Sirius Matters* (Dordrecht and London: Springer).

Buchner, E. (1982) *Die Sonnenuhr des Augustus* (Mainz: Philipp von Zabern).

Butrica, J.L. (1993) 'Propertius' Horoscope and a Birthdate Rejected', *CP* 88: 330–1.

Calcante, C.M. (2002) *Miracula rerum: Strategie semiologische del genere didascalico negli Astronomica di Manilio* (Pisa: Ed. ETS).

Callebat, L. (1994) 'Rhétorique et architecture dans le *De Architectura* de Vitruve', in P. Gros (ed.), *Le Projet de Vitruve: Objet, Destinataires et Réception du De Architectura* (Rome: L'École Française), 31–46.

Cardauns, B. (1976) *M. Terentius Varro: Antiquitates Rerum Divinarum* (Wiesbaden: Franz Steiner Verlag).

Chiabò, M. and Roberti, L. (1990) *Index Verborum Hygini De Astronomia* (Hildesheim and New York: Olms-Weidmann).

Cicu, L. (1979) 'La data dei *Phaenomena* di Germanico', *Maia* 31: 139–44.

Classen, C.J. (1968) 'Poetry and Rhetoric in Lucretius', *TAPA* 99: 77–118.

Clay, D. (1983) *Lucretius and Epicurus* (Ithaca and London: Cornell University Press).

Colborn, R. (2013) 'Solving Problems with Acrostics: Manilius dates Germanicus', *CQ* 63: 450–2

Cole, S. (2006) 'Cicero, Ennius, and the Concept of Apotheosis at Rome', *Arethusa* 39: 531–48.

Coleman, K.M. (1983) 'Manilius' Monster', *Hermes* 111: 226–32.

Coleman, R. (1977) *Virgil: Eclogues* (Cambridge: Cambridge University Press).

Connors, C. (1998) *Petronius the Poet: Verse and Literary Tradition in the Satyricon* (Cambridge: Cambridge University Press).

Cooley, A.E. (2009) *Res Gestae Divi Augusti: Text, Translation, and Commentary* (Cambridge: Cambridge University Press).

Courtney, E. (1969) 'Three Poems of Propertius', *BICS* 16: 70–87.

——— (1993) *The Fragmentary Latin Poets* (Oxford: Oxford University Press).

Coutts, R.L. (1964) 'Astrology in Horace, *Odes* II, 17', *PCA* 61: 26–7.

Cramer, F.H. (1951) 'Expulsion of Astrologers from Ancient Rome', *Classica et Mediaevalia* 12: 9–50.

——— (1954) *Astrology in Roman Law and Politics* (Philadelphia: American Philosophical Society).

Cumont, F. (1922) *After Life in Roman Paganism* (New Haven and London: Yale University Press).

Cuomo, S. (2011) 'Skills and Virtues in Vitruvius' Book 10', in M. Formisano and H. Böhme (eds), *War in Words: Transformations of War from Antiquity to Clausewitz* (Berlin and New York: De Gruyter), 309–32.

Debrohun, J.B. (2003) *Roman Propertius and the Reinvention of Elegy* (Ann Arbor: University of Michigan Press).

DeNardis, V.S. (2003) *Ratio omnia vincit: Cosmological, Political and Poetic Power in the Astronomica of Manilius* (unpublished diss., New York University).

Denyer, N. (1985) 'The Case Against Divination: An Examination of Cicero's *De Divinatione*', *PCPS* 31: 1–10.

de Vreese, J.G.W.M. (1927) *Petron 39 und die Astrologie* (Amsterdam: H.J. Paris).

DeWitt, N.W. (1930) 'Vergil and the Tragic Drama', *CJ* 26: 19–27.

Dick, B.F. (1963) 'The Technique of Prophecy in Lucan', *TAPA* 94: 37–49.
Dicks, D.R. (1963) 'Astrology and Astronomy in Horace', *Hermes* 91: 60–73.
Dieterich, A. (1900) 'Die Widmungselegie des letzten Buches des Propertius', *RhM* 55: 191–221.
Dirks, N.B., Eley, G. and Ortner, S.B. (eds) (1994) *Culture/Power/History: A Reader in Contemporary Social Theory* (Princeton: Princeton University Press).
Domenicucci, P. (1996) *Astra Caesarum: Astronomia, astrologia e catasterismo da Cesare a Domiziano* (Pisa: Ed. ETS).
Downing, E. (1990) 'Anti-Pygmalion: the *Praeceptor* in *Ars Amatoria*, Book 3', *Helios* 17: 237–49.
Due, O.S. (1974) *Changing Forms: Studies in the Metamorphoses of Ovid* (Copenhagen: Gyldendal).
Duff, J.W. and Duff, A.M. (1934) *Minor Latin Poets* (Cambridge, MA: Harvard University Press).
Durand, R. (1903) 'La date du *De Divinatione*', *Mélanges Boissier* (Paris: Albert Fontenoing), 173–83.
Durling, R.M. (1958) 'Ovid as *Praeceptor Amoris*', *CJ* 53: 157–67.
Dwyer, E.J. (1973) 'Augustus and the Capricorn', *Mitteilungen des Deutsches Archäologischen Instituts, Römische Abteilung* 80: 59–67.
Effe, B. (1977) *Dichtung und Lehre: Untersuchungen zur Typologie des antiken Lehrgedichts* (Munich: Beck).
Evans, J. (2004) 'The Astrologer's Apparatus: A Picture of Professional Practice in Greco-Roman Egypt', *Journal for the History of Astronomy* 35: 1–44.
——— and Berggren, J.L. (2006) *Geminos's Introduction to the Phenomena: A Translation and Study of a Hellenistic Survey of Astronomy* (Princeton: Princeton University Press).
Fairweather, J. (1981) *Seneca the Elder* (Cambridge: Cambridge University Press).
Fantham, E. (1985) 'Ovid, Germanicus and the Composition of the *Fasti*', *PLLS* 5: 243–81.
——— (2009) 'Caesar as an Intellectual', in M. Griffin (ed.), *A Companion to Julius Caesar* (Chichester: Wiley-Blackwell), 141–56.
Feeney, D. (1991) *The Gods in Epic* (Oxford: Clarendon Press).
——— (1992) '*Si licet et fas est*: Ovid's *Fasti* and the Problem of Free Speech under the Principate', in A. Powell (ed.), *Roman Poetry and Propaganda in the Age of Augustus* (London: Bristol Classical Press), 1–25.
——— (1998) *Literature and Religion at Rome* (Cambridge: Cambridge University Press).
——— (2007) *Caesar's Calendar: Ancient Time and the Beginnings of History* (Berkeley, CA and London: University of California Press).
Feraboli, S., Flores, E. and Scarcia, R. (1996–2001) *Manilio: Il poema degli astri (Astronomica)*, 2 vols (Rome: Mondadori).
Fishwick, D. (1992) 'The Statue of Julius Caesar in the Pantheon', *Latomus* 51: 329–36.

Fleury, P. (1990) *Vitruve De L'Architecture: Livre I* (Paris: Les Belles Lettres).

Flintoff, E. (1992) 'Comets and Confidence Tricks: A Meditation on Eclogue IX 47', *ACD* 28: 65–71.

Flower, M.A. (2008) *The Seer in Ancient Greece* (Berkeley, CA and London: University of California Press).

Fowler, D.P. (2000) 'The Didactic Plot', in M. Depew and D. Obbink (eds), *Matrices of Genre: Authors, Canons, and Society* (Cambridge, MA and London: Harvard University Press), 205–19.

Fox, M. (2004) 'Stars in the *Fasti*: Ideler (1825) and Ovid's Astronomy Revisited', *AJP* 125: 91–133.

Fraccaro, P. (1907) *Studi Varroniani: De Gente Populi Romani Libri IV* (Padua: Draghi).

Fraenkel, E. (1942) 'The Stars in the Prologue of the *Rudens*', *CQ* 36: 10–14.

Gain, D.B. (1976) *The Aratus Ascribed to Germanicus Caesar* (London: Athlone Press).

Gale, M.R. (2011) 'Digressions, Intertextuality, and Ideology in Didactic Poetry: The Case of Manilius', in S.J. Green and K. Volk (eds), *Forgotten Stars: Rediscovering Manilius' Astronomica* (Oxford: Oxford University Press), 205–21.

Galinsky, K. (1996) *Augustan Culture: An Interpretive Introduction* (Princeton: Princeton University Press).

Garrod, H.W. (1911) *Manili Astronomicon Liber II* (Oxford: Oxford University Press).

Gee E. (2000) *Ovid, Aratus, and Augustus: Astronomy in Ovid's Fasti* (Cambridge: Cambridge University Press).

——— (2001) 'Cicero's Astronomy', *CQ* 51: 520–36.

——— (2007) 'Quintus Cicero's Astronomy?', *CQ* 57: 565–85.

Getty, R.J. (1941) 'The Astrology of P. Nigidius Figulus (Lucan I, 649–65)', *CQ* 35: 17–22.

——— (1948) 'Some Astronomical Cruces in the *Georgics*', *TAPA* 79: 24–45.

——— (1960) 'Neopythagoreanism and Mathematical Symmetry in Lucan, *De Bello Civili* 1', *TAPA* 91: 310–23.

Gibson, R.K. (1997) 'Didactic Poetry as "Popular" Form: A Study of Imperatival Expressions in Latin Didactic Verse and Prose', in C. Atherton (ed.), *Form and Content in Didactic Poetry* (Bari: Levante), 67–98.

——— (2003) *Ovid: Ars Amatoria Book 3* (Cambridge: Cambridge University Press).

Gleason, M.W. (1995) *Making Men: Sophists and Self-Presentation in Ancient Rome* (Princeton: Princeton University Press).

Glucker, J. (1988) 'Cicero's Philosophical Affiliations', in J.M. Dillon and A.A. Long (eds), *The Question of Eclecticism* (Berkeley and London: University of California Press), 34–69.

Goold, G.P. (1954) 'De fonte codicum Manilianorum', *RhM* 97: 359–72.

Goold, G.P. (1959) 'Perseus and Andromeda: A Myth from the Skies', *Proceedings of the African Classical Associations* 11: 10–15.
——— (1961) 'A Greek Professorial Circle at Rome', *TAPA* 92: 168–92.
——— (1983) 'The Great Lacuna in Manilius', *Proceedings of the African Classical Associations* 17: 64–8.
——— (1985) *M. Manilii Astronomica* (Stuttgart: Teubner).
——— (1990) *Propertius Elegies* (Cambridge, MA and London: Harvard University Press).
——— (1997) *Manilius Astronomica* (Cambridge, MA and London: Harvard University Press).
Gradel, I. (2002) *Emperor Worship and Roman Religion* (Oxford: Oxford University Press).
Grafton, A.T. and Swerdlow, N.M. (1986) 'The Horoscope of the Foundation of Rome', *CP* 81: 148–53.
Green, S.J. (2001) '*Docens poeta*: Development of the Interviewer's Skills in Ovid's *Fasti*', *Latomus* 60: 603–12.
——— (2004) *Ovid Fasti I: A Commentary* (Leiden: Brill).
——— (2006) 'Lessons in Love: Fifty Years of Scholarship on Ovid's *Ars Amatoria* and *Remedia Amoris*', in R.K. Gibson, S.J. Green and A.R. Sharrock (eds), *The Art of Love: Bimillennial Essays on Ovid's Ars Amatoria and Remedia Amoris* (Oxford: Oxford University Press), 1–20.
——— (2008) 'Save our Cows? Augustan Discourse and Animal Sacrifice in Ovid's *Fasti*', *G&R* 55: 39–54.
——— (2009) 'Malevolent Gods and Promethean Birds: Contesting Augury in Augustus' Rome', *TAPA* 139: 147–67.
——— (2011) '*Arduum ad astra*: The Poetics and Politics of Horoscopic Failure in Manilius' *Astronomica*', in S.J. Green and K. Volk (eds), *Forgotten Stars: Rediscovering Manilius' Astronomica* (Oxford: Oxford University Press), 120–38.
——— and Volk, K. (eds) (2011) *Forgotten Stars: Rediscovering Manilius' Astronomica* (Oxford: Oxford University Press).
Griffin, M. (1972) 'The Elder Seneca and Spain', *JRS* 62: 1–19.
Grondona, M. (1980) *La Religione e la Superstizione nella Cena Trimalchionis* (Brussels: Latomus).
Gros, P. (1994) '*Munus non ingratum*: Le traité vitruvien et la notion de service', in P. Gros (ed.), *Le Projet de Vitruve: Objet, Destinataires et Réception du De Architectura* (Rome: L'École Française), 75–90.
Guillaumont, F. (1984) *Philosophe et Augure: rescherches sur la théorie cicéronienne de la divination* (Brussels: Latomus).
Gurval, R.A. (1997) 'Caesar's Comet: The Politics and Poetics of an Augustan Myth', *MAAR* 42: 39–71.
Habinek, T.N. (2007) 'Probing the Entrails of the Universe: Astrology as Bodily Knowledge in Manilius' *Astronomica*', in J. König and T. Whitmarsh

(eds), *Ordering Knowledge in the Roman Empire* (Cambridge: Cambridge University Press), 229–40.

Håkanson, L. (1989) *L. Annaeus Seneca Maior, Oratorum et Rhetorum Sententiae, Divisiones, Colores* (Leipzig: Teubner).

Hannah, R. (1996) 'Lucan *Bellum Civile* 1.649–65: The Astrology of P. Nigidius Figulus Revisited', *PLLS* 9: 175–90.

——— (2005) *Greek and Roman Calendars: Constructions of Time in the Classical World* (London: Duckworth).

Hardie, P. (1997) 'Questions of Authority: The Invention of Tradition in Ovid *Metamorphoses* 15', in T. Habinek and A. Schiesaro (eds), *The Roman Cultural Revolution* (Cambridge: Cambridge University Press), 182–98.

Harris, W.V. (1989) *Ancient Literacy* (Cambridge, MA and London: Harvard University Press).

Heath, M. (1985) 'Hesiod's Didactic Poetry', *CQ* 35: 245–63.

Heinze, R. (1919) *Ovids elegische Erzählung* (Leipzig: Teubner).

Herbert-Brown, G. (1994) *Ovid and the Fasti: An Historical Study* (Oxford: Clarendon Press).

——— (2002) 'Ovid and the Stellar Calendar', in G. Herbert-Brown (ed.), *Ovid's Fasti: Historical Readings at its Bimillenium* (Oxford: Oxford University Press), 101–28.

Heslin, P.J. (2007) 'Augustus, Domitian and the so-called Horologium Augusti', *JRS* 97: 1–20.

Heubeck, A., West, S. and Hainsworth, J.B. (1988) *A Commentary on Homer's Odyssey Volume 1: Introduction and Books I–VIII* (Oxford: Clarendon Press).

Heyworth, S.J. (2007a) *Sexti Properti Elegos* (Oxford: Oxford University Press).

——— (2007b) *Cynthia: A Companion to the Text of Propertius* (Oxford: Oxford University Press).

Hinds, S. (1987) *The Metamorphosis of Persephone: Ovid and the Self-Conscious Muse* (Cambridge: Cambridge University Press).

——— (1998) *Allusion and Intertext: Dynamics of Appropriation in Roman Poetry* (Cambridge: Cambridge University Press).

Holford-Strevens, L. (1997) 'Favorinus: the Man of Paradoxes', in J. Barnes and M. Griffin (eds), *Philosophia Togata II: Plato and Aristotle at Rome* (Oxford: Clarendon Press), 188–217.

Holleman, A.W.J. (1969) 'Ovidii *Metamorphoseon* liber XV 622–870, (*Carmen et Error*?)', *Latomus* 28: 42–60.

——— (1971) 'Ovid and Politics', *Historia* 20: 458–66.

Hopkinson, N. (2000) *Ovid Metamorphoses Book XIII* (Cambridge: Cambridge University Press).

Horsfall, N. (1972) 'Varro and Caesar: Three Chronological Problems', *BICS* 19: 120–8.

Horsfall, N. (1997) 'Criteria for the Dating of Calpurnius Siculus', *RFIC* 125: 166–96.

——— (2006) *Virgil, Aeneid 3: A Commentary* (Leiden: Brill).

Housman, A.E. (1903–30) *M. Manilii Astronomicon Libri*, 5 vols (London: The Richards Press).

Howe, T.N. and Rowland, I.D. (1999) *Vitruvius: Ten Books on Architecture* (Cambridge: Cambridge University Press).

Hübner, W. (1984) 'Manilius als Astrologe und Dichter', *ANRW* 2.32.1: 126–320.

——— (2011) 'Tropes and Figures: Manilian Style as a Reflection of Astrological Tradition', in S.J. Green and K. Volk (eds), *Forgotten Stars: Rediscovering Manilius' Astronomica* (Oxford: Oxford University Press), 141–64.

Hunter, R.L. (1995) 'Written in the Stars: Poetry and Philosophy in the *Phaenomena* of Aratus', *Arachnion* 2.

Hutchinson, G.O. (2002) 'The Publication and Individuality of Horace's *Odes* Books 1-3', *CQ* 52: 517–37.

——— (2006) *Propertius Elegies Book IV* (Cambridge: Cambridge University Press).

——— (2008) *Talking Books: Readings in Hellenistic and Roman Books of Poetry* (Oxford: Oxford University Press).

Ideler, J. (1825) 'Über den Astronomischen Teil der *Fasti* des Ovid', *Abhandlungen der königlichen Akademie der Wissenschaften zu Berlin 1822–3* (Berlin), 137–69.

Ihm, M. (1964) *C. Suetoni Tranquilli Opera, Vol. 1, De Vita Caesarum Libri VIII* (Stuttgart: Teubner).

Jeppesen, K.K. (1994) 'The Identity of the Missing *Togatus* and other Clues to the Interpretation of the Gemma Augustea', *OJA* 13: 335–55.

Jocelyn, H.D. (1967) *The Tragedies of Ennius: The Fragments* (Cambridge: Cambridge University Press).

Jones, A. (1999) *Astronomical Papyri from Oxyrhynchus* (Philadelphia: American Philosophical Press).

——— (2007) 'Astrologers and their Astronomy', in A.K. Bowman, R.A. Coles, N. Gonis, D. Obbink and P.J. Parsons (eds), *Oxyrhynchus: A City and its Texts* (London: Egypt Exploration Society), 307–14.

Jones, F. (1984) 'A Note of Manilius 5.604', *Acta Classica* 27: 139.

Joost-Gaugier, C.L. (1998) 'The Iconography of Sacred Space: A Suggested Reading of the Meaning of the Roman Pantheon', *Artibus et Historiae* 19.38: 21–42.

Kaufhold, S.D. (1996) 'A Look *Back* at Propertius 4.1.150', *Eranos* 94: 29–33.

Kennedy, D.F. (1993) *The Arts of Love: Five Studies in the Discourse of Roman Love Elegy* (Cambridge: Cambridge University Press).

——— (2011) 'Sums in Verse or a Mathematical Aesthetic?', in S.J. Green and K. Volk (eds), *Forgotten Stars: Rediscovering Manilius' Astronomica* (Oxford: Oxford University Press), 165–87.

Kenney, E.J. (1961) *P. Ovidi Nasonis: Amores, Medicamina Faciei Femineae, Ars Amatoria, Remedia Amoris* (Oxford: Clarendon Press).

Keppie, L. (1984) *The Making of the Roman Army: From Republic to Empire* (London: Batsford).

Keyser, P.T. (1992) 'Propertius' Horoscope: A Suggested Birthdate', *CP* 87: 328–34.

Kidd, D.A. (1982) 'Astrology for Maecenas', *Antichthon* 16: 88–96.

——— (1997) *Aratus: Phaenomena* (Cambridge: Cambridge University Press).

Kleiner, D.E.E. (1992) *Roman Sculpture* (New Haven and London: Yale University Press).

Knox, P.E. (1986) *Ovid's Metamorphoses and the Traditions of Augustan Poetry* (Cambridge: Cambridge Philological Society).

König, A. (2007) 'Knowledge and Power in Frontinus' *On Aqueducts*', in J. König and T. Whitmarsh (eds), *Ordering Knowledge in the Roman Empire* (Cambridge: Cambridge University Press), 177–205.

——— (2009) 'From Architect to Imperator: Vitruvius and his Addressee in the *De Architectura*', in L. Taub and A. Doody (eds), *Authorial Voices in Greco-Roman Technical Writing* (Trier: Wissenschatlicher Verlag Trier), 31–52.

König, J. (2007) 'Ordering Knowledge', in J. König and T. Whitmarsh (eds), *Ordering Knowledge in the Roman Empire* (Cambridge: Cambridge University Press), 3–39.

——— and Whitmarsh, T. (eds) (2007) *Ordering Knowledge in the Roman Empire* (Cambridge: Cambridge University Press).

Kraft, K. (1967) 'Zum Capricorn auf den Münzen des Augustus', *Jahrbuch für Numismatik und Geldgeschichte* 17: 17–27.

Krappe, A.H. (1927) 'Tiberius and Thrasyllus', *AJP* 48: 359–66.

Kroll, W. and Skutsch, F. (1897–1913) *Iulii Firmici Materni Matheseos Libri VIII* (Leipzig: Teubner).

Kronk, G.W. (1999) *Cometography: A Catalog of Comets, I: Ancient to 1799* (Cambridge: Cambridge University Press).

Krostenko, B.A. (2000) 'Beyond (Dis)belief: Rhetorical Form and Religious Symbol in Cicero's *de Divinatione*', *TAPA* 130: 353–91.

Kubiak, D.P. (1994) 'Aratean Influence in the *De Consulatu Suo* of Cicero', *Philologus* 138: 52–66.

Lacey, W.K. (1978) *Cicero and the End of the Republic* (New York: Barnes & Noble).

Lancaster, L.C. (2005) *Concrete Vaulted Construction in Imperial Rome: Innovations in Context* (Cambridge: Cambridge University Press).

Landolfi, L. (1991) 'Manilio e la gerarchia delle stelle (*Astr.* 5.734-45)', *Prometheus* 17: 247–58.

Le Boeuffle, A. (1965) 'Recherches sur Hygin', *REL* 43: 275–94.

Le Boeuffle, A. (1975) *Germanicus: Les Phénomènes d'Aratos* (Paris: Les Belles Lettres).
——— (1977) *Les Noms Latins d'Astres et de Constellations* (Paris: Les Belles Lettres).
——— (1983a) *Hygin L'Astronomie* (Paris: Les Belles Lettres).
——— (1983b) 'Le destin astral d'après Germanicus, auteur des *Phénomènes*', in J. Duchemin (ed.), *Visage du destin dans les mythologies: mélanges Jacqueline Duchemin: Actes du Colloque de Chantilly, 1–2 Mai 1980* (Paris: Les Belles Lettres), 87–93.
——— (1989) *Le Ciel des Romains* (Paris: De Boccard).
Le Bonniec, H. and Le Boeuffle, A. (1972) *Pline L'Ancien: Histoire Naturelle XVIII* (Paris: Les Belles Lettres).
Lehoux, D. (2007) *Astronomy, Weather, and Calendars in the Ancient World: Parapegmata and Related Texts in Classical and Near-Eastern Societies* (Cambridge: Cambridge University Press).
——— (2012) *What Did the Romans Know? An Inquiry into Science and Worldmaking* (Chicago, IL and London: University of Chicago Press).
Levene, D.S. (2012) 'Defining the Divine in Rome: In memoriam S.R.F. Price', *TAPA* 142: 41–81.
Lewis, A.-M. (1986) 'Rearrangement of Motif in Latin Literature: The Emergence of a Roman *Phaenomena*', in C. Deroux (ed.), *Studies in Latin Literature and Roman History* 4 (Brussels: Latomus), 210–33.
——— (1992) 'The Popularity of the *Phaenomena* of Aratus: A Reevaluation', in C. Deroux (ed.), *Studies in Latin Literature and Roman History* 6 (Brussels: Latomus), 94–118.
——— (1998) 'What Dreadful Purpose Do We Have? A New Explanation for the Astrological Prophecy of Nigidius Figulus in Lucan's *Pharsalia* I, 658-63', in C. Deroux (ed.), *Studies in Latin Literature and Roman History* 9 (Brussels: Latomus), 379–400.
——— (2008) 'Augustus and his Horoscope Reconsidered', *Phoenix* 62: 308–37.
Liebeschuetz, J.H.W.G. (1979) *Continuity and Change in Roman Religion* (Oxford: Clarendon Press).
Linderski, J. (1982) 'Cicero and Roman Divination', *La Parola del Passato* 37: 12–38.
Lindsay, W.M. (1913) *Sexti Pompei Festi De verborum significatu quae supersunt cum Pauli epitome* (Leipzig: Teubner).
Lintott, A. (2008) *Cicero as Evidence: A Historian's Companion* (Oxford: Oxford University Press).
Loercke, W.C. (1990) 'A Rereading of the Interior Elevation of Hadrian's Rotunda', *Journal of the Society of Architectural Historians* 49: 22–43.
Long, A.A. (1982) 'Astrology: Arguments Pro and Contra', in J. Barnes, J. Brunschwig, M. Burnyeat and M. Schofield (eds), *Science and Speculation:*

Studies in Hellenistic Theory and Practice (Cambridge: Cambridge University Press), 165–92.
Lowe, D. (2004) 'Monsters in the Roman Sky: Heaven and Earth in Manilius' *Astronomica*', in P.L. Yoder and P.M. Kreuter (eds), *Monsters and the Monstrous: Myths and Metaphors of Enduring Evil* (Interdisciplinary Press), 143–56.
Lyne, R.O.A.M. (1995) *Horace: Behind the Public Poetry* (New Haven, CT: Yale University Press).
McCartney, E.S. (1926) 'The Classical Astral Weather Chart for Rustics and for Seamen', *Classical Weekly* 20: 43–9, 51–4.
McDermott, E.A. (1982) 'Horace, Maecenas and *Odes* 2.17', *Hermes* 110: 211–28.
MacDonald, W.L. (1976) *The Pantheon: Design, Meaning and Progeny* (London: Allen Lane).
McEwen, I.K. (2003) *Vitruvius: Writing the Body of Architecture* (Cambridge, MA and London: MIT Press).
Macleod, C.W. (1976) 'Propertius 4.1', *PLLS* 1: 141–53.
MacMullen, R. (1967) *Enemies of the Roman Order: Treason, Unrest, and Alienation in the Empire* (Cambridge, MA: Harvard University Press).
——— (1971) 'Social History in Astrology', *Ancient Society* 2: 105–16.
Maltby, R. (1991) *A Lexicon of Ancient Latin Etymologies* (Leeds: Cairns).
Mann, W. (2011) 'On Two Stoic "Paradoxes" in Manilius', in S.J. Green and K. Volk (eds), *Forgotten Stars: Rediscovering Manilius' Astronomica* (Oxford: Oxford University Press), 85–103.
Marquis, E. (1973) 'Propertius 4.1.150 and the Gate of Cancer', *WS* 86: 126–33.
Marshall, P.K. (1990) *A. Gellii Noctes Atticae* (Oxford: Clarendon Press).
Martin, Joseph (1953) *T. Lucreti Cari De Rerum Natura Libri Sex* (Leipzig: Teubner).
Martin, Jean (1998) *Aratos Phénomènes* (Paris: Les Belles Lettres).
Masterton, M. (2004) 'Status, Pay, and Pleasure in the *De Architectura* of Vitruvius', *AJP* 125: 387–416.
May, J.M. (2002) 'Cicero: His Life and Career', in J.M. May (ed.), *Brill's Companion to Cicero* (Leiden: Brill), 1–21.
Mayhoff, C. (1906) *C Plini Secundi Naturalis Historiae Libri I-VI* (Leipzig: Teubner).
Miller, J.F. (1993) 'Apostrophe, Aside and the Didactic Addressee: Poetic Strategies in *Ars Amatoria* III', *MD* 31: 231–41.
Mitsis, P. (1993) 'Committing Philosophy on the Reader: Didactic Coercion and Reader Autonomy in *De Rerum Natura*', *MD* 31: 111–28.
Moatti, C. (1997) *La Raison de Rome: Naissance de l'esprit critique à la fin de la République* (Paris: Seuil).
Momigliano, A. (1984) 'The Theological Efforts of the Roman Upper Classes in the First Century B.C.', *CP* 79: 199–211.

Montanari Caldini, R. (1973) 'L'astrologia nei Prognostica di Germanico', *SIFC* 45: 137–204.
—— (1976) 'L'astrologia nella traduzione aratea di Germanico', *SIFC* 48: 29–117.
—— (1979) *Horos e Properzio: ovvero l'ispirazione necessaria* (Florence: Clusf).
Morford, M.P.O. (1967) *The Poet Lucan: Studies in Rhetorical Epic* (Oxford: Blackwell).
—— (2002) *The Roman Philosophers: From the Time of Cato the Censor to the Death of Marcus Aurelius* (London: Routledge).
Mueller, K. (1995) *Petronii Arbitri Satyricon Reliquiae* (Stuttgart: Teubner).
Murgatroyd, P. (1994) 'Narrative Techniques in Manilius, *Astronomica* 5, 538–618', in C. Deroux (ed.), *Studies in Latin Literature and Roman History* 7 (Brussels: Latomus), 416–29.
Murgia, C.E. (1989) 'Propertius 4.1.87–88 and the Division of 4.1', *HSCP* 92: 257–72.
Myerowitz, M. (1985) *Ovid's Games of Love* (Detroit: Wayne State University Press).
Myers, K.S. (2009) *Ovid Metamorphoses Book XIV* (Cambridge: Cambridge University Press).
Mynors, R.A.B. (1969) *P. Vergili Maronis Opera* (Oxford: Clarendon Press).
Nethercut, W.R. (1968) 'Elegiac Technique in Propertius 4.1.71-150', *WS* 81: 92–7.
—— (1970) 'The Astrological Significance of Propertius IV.1.150', *WS* 83: 110–17.
Neuburg, M. (1993) 'Hitch Your Wagon to a Star: Manilius and his Two Addressees', *MD* 31: 243–82.
Neugebauer, O. and Van Hoesen, H.B. (1959) *Greek Horoscopes* (Philadelphia: American Philosophical Society).
Newlands, C.E. (1995) *Playing with Time: Ovid and the Fasti* (Ithaca: Cornell University Press).
Nice, A.T. (2001) 'Ennius or Cicero? The Disreputable Diviners at Cicero *De Divinatione* 1.132', *A Class.* 44: 153–66.
Nichols, M.F. (2009) 'Social Status and the Authorial Personae of Horace and Vitruvius', in L.B.T. Houghton and M. Wyke (eds), *Perceptions of Horace: A Roman Poet and His Readers* (Cambridge: Cambridge University Press), 109–22.
Nisbet, R.G. and Hubbard, M. (1970) *A Commentary on Horace Odes, Book 1* (Oxford: Clarendon Press).
—— (1978) *A Commentary on Horace Odes, Book 2* (Oxford: Clarendon Press).
—— and Rudd, N. (2004) *A Commentary on Horace Odes, Book 3* (Oxford: Oxford University Press).
North, J. (1990) 'Diviners and Divination at Rome', in M. Beard and J. North (eds), *Pagan Priests: Religion and Power in the Ancient World* (London: Duckworth), 49–71.

Ogden, D. (2008) *Perseus* (London: Routledge).

Oliver, R.P. (1980) 'Thrasyllus in Tacitus (*Ann.* 6.21)', *ICS* 5: 130–48.

Panayotakis, C. (2010) *Decimus Laberius: The Fragments* (Cambridge: Cambridge University Press).

Pavis d'Escurac, H. (1981) 'La pratique augurale romaine à la fin de la République: Scepticisme et Tradition', in *Religion et Culture dans la Cité Italienne de l'Antiquité à Nos Jours* (Strasbourg: University of Strasbourg), 27–39.

Pease, A.S. (1920–3) *M. Tulli Ciceronis De Divinatione Libri Duo* (Urbana: University of Illinois).

Pendergraft, M.L.B. (1990) 'On the Nature of the Constellations: Aratus, *Ph.* 367-85', *Eranos* 88: 99–106.

Perret, J. (1983) 'L'Exaltation de Daphnis (Virgile, *Buc.* V)', in *Hommages à Jean Cousin: Rencontres avec L'Antiquité Classique* (Paris: Les Belles Lettres), 123–32.

Pollini, J. (1993) 'The Gemma Augustea: Ideology, Rhetorical Imagery, and the Creation of a Dynastic Narrative', in P.J. Holliday (ed.), *Narrative and Event in Ancient Art* (Cambridge: Cambridge University Press), 258–98.

Possanza, D.M. (1992) 'Two Notes on Q. Cicero's *De Duodecim Signis*', *CP* 87: 44–6.

——— (2004) *Translating the Heavens: Aratus, Germanicus, and the Poetics of Latin Translation* (New York: Peter Lang).

Potter, D.S. (1994) *Prophets and Emperors: Human and Divine Authority from Augustus to Theodosius* (Cambridge, MA and London: Harvard University Press).

Purcell, N. (1983) 'The *Apparitores*: A Study in Social Mobility', *Papers of the British School at Rome* 51: 123–73.

Ramage, E.S. (1985) 'Augustus' Treatment of Julius Caesar', *Historia* 34: 223–45.

Ramsey, J.T. (2003) *Cicero: Philippics I-II* (Cambridge: Cambridge University Press).

——— (2007) 'A Catalogue of Greco-Roman Comets from 500 B.C. to A.D. 400', *JHA* 38: 175–97.

——— and Licht, A.L. (1997) *The Comet of 44 B.C. and Caesar's Funeral Games* (Atlanta: Scholars Press).

Rawson, E. (1975) *Cicero: A Portrait* (London: Allen Lane).

——— (1978) 'Caesar, Etruria and the *Disciplina Etrusca*', *JRS* 68: 132–52.

——— (1985) *Intellectual Life in the Late Roman Republic* (London: Duckworth).

Reay, B. (2005) 'Agriculture, Writing, and Cato's Aristocratic Self-Fashioning', *Classical Antiquity* 24: 331–61.

Rehak, P. (2006) *Imperium and Cosmos: Augustus and the Northern Campus Martius* (Madison: University of Wisconsin Press).

Rimell, V. (2007) 'Petronius' Lessons in Learning—the Hard Way', in J. König and T. Whitmarsh (eds), *Ordering Knowledge in the Roman Empire* (Cambridge: Cambridge University Press), 108–32.

Ripat, P. (2011) 'Expelling Misconceptions: Astrologers at Rome', *CP* 106: 115–54.

Rose, H.J. (1927) 'Mox', *CQ* 21: 57–66.

Rose, K.F.C. and Sullivan, J.P. (1968) 'Trimalchio's Zodiac Dish (Petronius, *Sat.* 35.1-5)', *CQ* 18: 180–4.

Rudd, N. (1989) *Horace: Epistles Book II and Epistle to the Pisones ('Ars Poetica')* (Cambridge: Cambridge University Press).

Salzman, M.R. (1998) 'Deification in the *Fasti* and the *Metamorphoses*', in C. Deroux (ed.), *Studies in Latin Literature and Roman History* 9 (Brussels: Latomus), 313–46.

Sandbach, F.H. (1962) 'Some Problems in Propertius', *CQ* 12: 263–76.

Santini, C. (1977) *Il segno e la tradizione in Germanico scrittore* (Rome: Cadmo).

Schmeling, G. (2011) *A Commentary on the Satyrica of Petronius* (Oxford: Oxford University Press).

Schofield, M. (1986) 'Cicero For and Against Divination', *JRS* 76: 47–65.

Schütz, M. (1990) 'Zur Sonnenuhr des Augustus auf dem Marsfeld', *Gymnasium* 97: 432–57.

Scott, K. (1941) 'The *sidus Iulium* and the Apotheosis of Caesar', *CP* 36: 257–72.

Sedley, D.N. (1993) 'Chrysippus on Psychophysical Causality', in J. Brunschwig and M.C. Nussbaum (eds), *Passions and Perceptions* (Cambridge: Cambridge University Press), 313–31.

Segal, C. (1984) 'Senecan Baroque: The Death of Hippolytus in Seneca, Ovid, and Euripides', *TAPA* 114: 311–25.

Shackleton Bailey, D.R. (1985) *Q. Horati Flacci Opera* (Leipzig: Teubner).

Sharples, R.W. (1991) *Cicero: On Fate and Boethius: The Consolation of Philosophy IV.5–7, V* (Warminster: Aris & Phillips).

Sharrock, A.R. (1994a) *Seduction and Repetition in Ovid's Ars Amatoria 2* (Oxford: Clarendon Press).

——— (1994b) 'Ovid and the Politics of Reading', *MD* 33: 97–122.

——— (1997) 'Response by Alison Sharrock: *Haud mollia iussa*', in C. Atherton (ed.), *Form and Content in Didactic Poetry* (Bari: Levante), 99–115.

——— (2005) 'Those Who Can, Teach: Ovid's *Ars Amatoria* and Contemporary Instructional Writing', in M. Horster and C. Reitz (eds), *Wissensvermittlung in dichterischer Gestalt* (Stuttgart: F. Steiner), 243–63.

Shotter, D.C.A. (1972) 'The Trial of M. Scribonius Libo Drusus', *Historia* 21: 88–98.

Skutsch, O. (1985) *The Annals of Q. Ennius* (Oxford: Clarendon Press).

Smith, M.S. (1975) *Cena Trimalchionis Petroni Arbitri* (Oxford: Clarendon Press).

Sogno, C. (2005) 'Astrology, Morality, the Emperor, and the Law in Firmicus Maternus' *Mathesis*', *Illinois Classical Studies* 30: 167–76.

Soubiran, J. (1969) *Vitruve: De L'architecture: Livre IX* (Paris: Les Belles Lettres).

Soubiran, J. (1972) *Cicéron, Aratea: Fragments poétiques* (Paris: Les Belles Lettres).

Stahl, H.-P. (1985) *Propertius 'Love' and 'War': Individual and State under Augustus* (Berkeley, CA and London: University of California Press).

Steinby, E.M. (1993–2000) *Lexicon Topographicum Urbis Romae* (Rome: Quasar).

Strauss Clay, J. (1993) 'The Education of Perses: From *mega nepios* to *dion genos* and Back', *MD* 31: 23–33.

Sussman, L.A. (1978) *The Elder Seneca* (Leiden: Brill).

Swain, S. (1989) 'Favorinus and Hadrian', *ZPE* 79: 150–8.

Swan, P.M. (2004), *The Augustan Succession: An Historical Commentary on Cassius Dio's Roman History, Books 55–56 (9 B.C.–A.D. 14)* (Oxford: Oxford University Press).

Syme, R. (1939) *The Roman Revolution* (Oxford: Clarendon Press).

Syndikus, H.P. (1972) *Die Lyrik des Horaz* (Darmstadt: Wissenschaftliche Buchgesellschaft).

Taub, L. (2003) *Ancient Meteorology* (London: Routledge).

Taylor, L.R. (1934) 'Varro's *De Gente Populi Romani*', *CP* 29: 221–9.

Tester, S.J. (1987) *A History of Western Astrology* (Woodbridge: Boydell).

Thomas, E. (2004) 'From the Pantheon of the Gods to the Pantheon of Rome', in R. Wrigley and M. Craske (eds.), *Pantheons: Transformations of a Monumental Idea* (Aldershot: Ashgate), 11–34.

Thorndike, L. (1913) 'A Roman Astrologer as a Historical Source: Julius Firmicus Maternus', *CP* 8: 415–35.

Tissol, G. (2002) 'The House of Fame: Roman History and Augustan Politics in *Metamorphoses* 11-15', in B.W. Boyd (ed.), *Brill's Companion to Ovid* (Leiden: Brill), 305–35.

Toohey, P. (1996) *Epic Lessons: An Introduction to Ancient Didactic Poetry* (London: Routledge).

Uden, J. (2011) 'A Song from the Universal Chorus: The Perseus and Andromeda Epyllion', in S.J. Green and K. Volk (eds), *Forgotten Stars: Rediscovering Manilius' Astronomica* (Oxford: Oxford University Press), 235–52.

Vahlen, J. (1903) *Ennianae Poesis Reliquiae* (Leipzig: Teubner).

Viré, G. (1992) *Hygini De Astronomia* (Stuttgart: Teubner).

Volk, K. (2002) *The Poetics of Latin Didactic: Lucretius, Vergil, Ovid, Manilius* (Oxford: Oxford University Press).

——— (2009) *Manilius and his Intellectual Background* (Oxford: Oxford University Press).

——— (2010) 'Aratus', in J.J. Clauss and M. Cuypers (eds), *A Companion to Hellenistic Literature* (Oxford: Wiley-Blackwell), 197–210.

——— (2011a) 'Reading the Sky: A Century of Manilian Scholarship', in S.J. Green and K. Volk (eds), *Forgotten Stars: Rediscovering Manilius' Astronomica* (Oxford: Oxford University Press), 1–10.

Volk, K. (2011b) 'Manilian Self-Contradiction', in S.J. Green and K. Volk (eds), *Forgotten Stars: Rediscovering Manilius' Astronomica* (Oxford: Oxford University Press), 104–19.

Voss, B.R. (1972) 'Die Andromeda—Episode des Manilius', *Hermes* 100: 413–34.

Wallace-Hadrill, A. (1997) '*Mutatio morum*: The Idea of a Cultural Revolution', in T. Habinek and A. Schiesaro (eds), *The Roman Cultural Revolution* (Cambridge: Cambridge University Press), 3–22.

Ward, D.J. (1968) 'Weather Signs and Weather Magic: Some Ideas on Causality in Popular Belief', *Pacific Coast Philology* 3: 67–72.

Warden, J. (1980) *Fallax Opus: Poet and Reader in the Elegies of Propertius* (Toronto: University of Toronto Press).

Wardle, D. (1997) '"The Sainted Julius": Valerius Maximus and the Dictator', *CP* 92: 323–45.

——— (2006) *Cicero on Divination Book 1* (Oxford: Clarendon Press).

——— (2009) 'Caesar and Religion', in M. Griffin (ed.), *A Companion to Julius Caesar* (Chichester: Wiley-Blackwell), 100–11.

Warmington, E.H. (1936–67) *Remains of Old Latin* (Cambridge, MA and London: Harvard University Press).

Weinstock, S. (1971) *Divus Julius* (Oxford: Clarendon Press).

Welch, T.S. (2005) *The Elegiac Cityscape: Propertius and the Meaning of Roman Monuments* (Columbus: Ohio State University Press).

Wellesley, K. (1989) *Cornelii Taciti Libri qui supersunt; Tom.2, fasc.1, Historiarum Libri* (Stuttgart: Teubner).

West, D. (1991) '*Cur me querelis* (Horace, *Odes* 2.17)', *AJP* 112: 45–52.

——— (1993) 'On Serial Narration and on the Julian Star', *PVS* 21: 1–16.

White, P. (1988) 'Julius Caesar in Augustan Rome', *Phoenix* 42: 334–56.

Williams, R.D. (1964) 'The Sixth Book of the *Aeneid*', *G&R* 11: 48–63.

Wills, J. (1996) *Repetition in Latin Poetry: Figures of Allusion* (Oxford: Clarendon Press).

Wilson Jones, M. (2000) *Principles of Roman Architecture* (New Haven and London: Yale University Press).

Wilson, A.M. (1986) 'The Prologue to Manilius 1', *PLLS* 5: 283–98.

Wiltshire, S.F. (1974) '*Nunc Age*: Lucretius as Teacher', *CB* 50: 33–7.

Wray, D. (2001) 'Astrology in Ancient Rome: Poetry, Prophecy and Power', *Fathom Archive*, at <http://fathom.lib.uchicago.edu/1/777777122543/>.

Yon, A. (1944) *Cicéron: Traité du Destin* (Paris: Les Belles Lettres).

Zanker, P. (1988) *The Power of Images in the Age of Augustus* (Ann Arbor: University of Michigan Press).

Zehnacker, H. (1989) 'D'Aratos à Aviénus: Astronomie et Idéologie', *Illinois Classical Studies* 14: 317–29.

Zetzel, J.E.G. (1995) *Cicero De Re publica: Selections* (Cambridge: Cambridge University Press).

Index Locorum

Aeschylus
Agamemnon
966–7: 130

Apollonius Rhodius
2.516–17: 130

Appian
Civil Wars
2.135: 92

Aratus
5–14: 130
75: 15
264–7: 130, 136
283–6: 135
293: 130
313–15: 130, 135
329–35: 130, 135
423–39: 136
454–61: 27, 146
529–33: 113
545: 135
682: 137
740–57: 130
744–7: 130
768–71: 130
886: 130
936: 130

Aristotle
Meteorology
340b–341a: 133

On the Heavens
289a 19–35: 133

Arrian
Anabasis
7.16.5–17.6: 194

Augustine
De Civitate Dei
5.2–6: 84

Augustus
Res Gestae
13: 106–7

Caesar
De Bello Gallico
8 *praef.* 5–6: 91

Calpurnius Siculus
1.77–83: 160, 195

Cassius Dio
44.22.2–34.1: 92
45.1.3–5: 67
45.7.1–2: 70, 157, 160, 162
49.43.1–5: 103
53.27.2–3: 100–1
55.11.2–3: 105
56.25.5: 72–4, 103, 104
57.19.3–4: 106
58.27.1–3: 194
66.24.2: 101

Cato The Elder
De Re Rustica
1.5.4: 65–6

Cicero
Aratea
fr. 1: 136–7
fr. 19: 136–7
fr. 33.20: 136–7
fr. 33.37–41: 136
fr. 33.57–9: 134–5
fr. 33.89–90: 135, 142
fr. 33.107–19: 135
fr. 33.155–66: 137
fr. 33.188: 136–7
fr. 33.198–202: 135–6
fr. 33.294: 136–7
fr. 33.320: 135, 141
fr. 33.381–2: 49
fr. 33.416: 49

De Divinatione
1.2: 81, 82
1.3: 81
1.4: 81
1.8: 77
1.10–11: 77
1.13–15: 137
1.13: 79
1.17–22: 79
1.17–18: 76, 81
1.22: 79
1.25: 79
1.33: 79
1.36: 80, 81, 82
1.72: 68
1.91–3: 81, 82
1.106: 79
1.119: 70
1.132: 65, 82–3, 84
2.28–69: 77
2.28: 83
2.60: 91
2.70–4: 77
2.87–99: 6, 75, 80–6
2.88: 83
2.89: 84, 87
2.90–1: 83, 84, 88–9, 196
2.92–3: 83, 84, 192, 196
2.94: 84
2.95: 84, 192, 196
2.96–7: 84, 196
2.98: 84, 85, 192, 196
2.99: 66–7, 69, 83, 84, 85, 192
2.150: 78, 79

De Fato
1–2: 86
2–3: 90–1
7–8: 75, 87–93
11–16: 89

De Haruspicum Responsis
18–20: 77

De Legibus
2.31–3: 77

De Natura Deorum
1.119: 154, 156
2.14: 66
2.62: 156

De Officiis
1.150–1: 111

De Oratore
1.69: 12
3.126–7: 111
3.132–6: 111
3.154: 156

De Re Publica
1.1–13: 156
1.21–2: 12, 65
2.4–20: 156
6.9–29: 156
6.14: 163
6.24: 163
6.26: 164

Epistulae ad Atticum
2.1.11: 134
14.12.2: 87

Epistulae ad Familiares
4.13.4: 81
6.6.7: 69

Epistulae ad Quintum Fratrem
fr. 3.5.1–2: 78

In Verrem
2.2.129: 69

Philippics
1–2: 87, 92

Pro Marcello
8: 156
27: 156

Tusculanae Disputationes
1.28: 156–7

Columella
11.1.32: 131

Diodorus Siculus
fr. 36.5.1: 66

Diomedes the Grammarian
1.482.21: 40

ENNIUS
Annales
1 fr. 54: 155–7, 167
1 fr. 110: 155–7

FIRMICUS MATERNUS
Mathesis
1.3: 173–4
2 *praef.* 1: 26
2.14.1: 26
2.20.14: 26
2.30.3–10: 60
3–4: 37
3.1.19: 26

GELLIUS
14.1: 7, 196–7
14.1.2: 196
14.1.7–11: 192
14.1.9: 131
14.1.33: 184
15.4.3: 70

GEMINOS
Introduction to the Phenomena
3.14: 132
17: 131–2

GERMANICUS: 143–9
1–725: 190
1–16: 143–4
96–139: 145
186: 145
250: 145
444–5: 146, 148
554–60: 99, 143, 144–5, 164
frr. 3–4: 146–8
fr. 3.10–22: 146–7
fr. 4.40–1: 147

HESIOD
Works and Days
564–7: 48
587–8: 130
619–20: 48

HOMER
Iliad
5.5–6: 49
18.607–8: 48
22.26–31: 130, 131

Odyssey
11.601–4: 169

HORACE
Epistles
1.10.16–17: 142
2.2.183–9: 122, 125

Epodes
1.27: 142
3.15–16: 142
15.7–8: 142
16.61–2: 142

Odes
1.11: 121–3
1.12.46: 158
1.17.17–18: 142
1.28.21–2: 142
2.17: 123-5
2.17.17–25: 123–4
2.17.19–20: 98
2.17.27–30: 124
3.1.27–8: 142
3.1.31–2: 142
3.3.15–16: 167
3.25.3–6: 158
3.29.17–20: 142
3.29.29–32: 122
4.14.21: 142

Satires
1.6.113–14: 83, 174
1.9: 187

HYGINUS
De Astronomia: 125–7
praef. 31–2: 127
1.8: 127
2.35: 130
4.5: 127

JULIUS OBSEQUENS
68: 162

JUVENAL
6.553–91: 123
6.582–91: 83, 174
10.94: 106

Livy
39.16.8: 83

Lucan
1.639–72: 67, 69

Lucretius
1.50: 22
1.130: 22
1.143–4: 42
1.146: 42
1.803–11: 40–1
1.897–903: 40–1
1.931–50: 29
1.1114–17: 42
2.55–61: 42–3
2.840: 22
3.87–93: 42–3
4.1–25: 29
4.912: 22
5.420: 22
6.35–41: 42–3

Lygdamus
3.5.2: 141

Manilius
1: 13–21, 22, 190
1.1–6: 13–14
1.1–2: 120
1.7–10: 14–15, 56, 59, 172
1.11–13: 14
1.15: 27
1.20–4: 16, 57
1.51–2: 27
1.120: 20
1.194–5: 14–16
1.215–17: 21
1.242: 15
1.255–6: 17
1.260: 17
1.346–7: 49
1.354–60: 46–7
1.373–81: 15
1.373–4: 17–19
1.396–411: 191
1.410–11: 48
1.458–60: 19
1.470–1: 53
1.474–7: 19–20
1.552: 17
1.557–60: 20
1.560A: 20
1.648–9: 20
1.670–1: 27
1.811: 20
1.859: 20
1.898–903: 11
1.904: 20

2: 16, 21–8
2.1–59: 12, 37
2.25–38: 191
2.83: 21
2.86: 21
2.105–8: 21–2, 24, 43
2.115–25: 22–3
2.136–44: 23–4
2.145–9: 24–5
2.157–8: 22
2.203–5: 22
2.234–5: 22
2.270: 31
2.297–341: 22
2.331: 31
2.423: 20
2.433: 39
2.534: 20
2.643–4: 27, 31
2.693–712: 192
2.693: 22
2.738–48: 26
2.749–54: 25–6
2.755–60: 26
2.761–71: 26
2.772–83: 26
2.784–7: 26
2.788: 22
2.842: 22
2.856–967: 29
2.898–9: 22
2.927: 22
2.951: 20
2.961–7: 27

3: 28–33, 41
3.5–35: 12, 37, 56
3.31–42: 16, 28–9, 57
3.36–7: 29
3.43–159: 29
3.43: 29
3.46: 29
3.96–155: 27
3.142–4: 29–30

3.149–55: 29–30
3.156–9: 26, 32
3.160–202: 98
3.186: 29
3.203–509: 29, 30–1
3.203–4: 29, 39
3.211: 29
3.218–46: 30–1
3.276: 29
3.295: 29
3.389: 20
3.483–509: 30–1
3.506: 20
3.560–617: 31–2
3.581–9: 26, 27, 31–2
3.618: 29

4: 13, 33–44, 54
4.14–22: 30
4.57–62: 171
4.98–101: 30
4.122–293: 33–8, 42, 45
4.124–34: 36–7
4.140–7: 36
4.148–9: 37
4.157: 36
4.165–72: 36–7
4.186: 37
4.190: 36
4.192: 37
4.194: 37
4.197–9: 37
4.203–8: 36
4.209–12: 37
4.213–14: 37
4.217–22: 36–7
4.247: 37
4.272: 37
4.294–386: 38, 42
4.308–9: 38, 41–2
4.368: 38
4.380–6: 38–9
4.387–9: 39–42, 54
4.390–407: 42
4.408–501: 42
4.409: 31, 42
4.431–42: 16, 57
4.502–84: 42
4.542–6: 97
4.711–817: 192
4.769–77: 192
4.773–7: 97, 184
4.791–6: 98
4.866–72: 42–3, 54
4.932: 43
4.935: 14

5: 44–53
5.1–7: 44–5
5.8–11: 45
5.27–9: 52
5.32–709: 45–52
5.416–17: 49
5.504–8: 51
5.516–19: 51
5.538–618: 46–52
5.551: 51
5.569–72: 50
5.581–6: 47
5.592–4: 47
5.595–7: 47
5.598–600: 47
5.604: 49
5.608–11: 47–8
5.612–15: 48, 49–50
5.619–30: 50–2
5.627–8: 50–1
5.709: 44, 52
5.710–45: 52–3
5.740–5: 53

MARTIAL
5.56.10–11: 111

OVID
Amores
2.16.3–4: 142
3.8.51–2: 158

Ars Amatoria
1.1: 15
1.31–4: 61
2.599–600: 61
3.1–6: 57–8
3.7–8: 40
3.43–4: 58–9
3.57–8: 61
3.281: 58–9
3.353–4: 58–9
3.613–14: 61
3.667–8: 58–9

Epistulae Ex Ponto
4.13.25–6: 164

Fasti
1.2: 190
1.49–52: 19
1.295–310: 127–8
1.295–6: 128, 190
1.310: 128
1.315–16: 191
2.144: 154
2.487: 155
2.496: 167
3.55–8: 27
3.155–60: 171
3.199–200: 27
3.697–702: 168–9
4.901–4: 191
4.939–40: 142
5.147–8: 27
6.277–80: 65

Metamorphoses
4.539–42: 166
4.706–8: 48
4.720: 47
4.721–2: 48
4.727: 47
4.734: 47
4.740: 49
9.262–72: 164–5
13.949–55: 166
14.600–8: 165–6
14.814: 155
14.818–26: 167
14.824–8: 166–7
14.845–51: 167–8
15.492–546: 169–70
15.498: 169
15.525: 169
15.547–8: 170
15.626–744: 153
15.746–851: 151–4
15.746–61: 152–3
15.752–8: 171–2
15.760: 153
15.766: 169
15.767: 153
15.776: 169
15.777–8: 169
15.787: 160
15.840–51: 163–4
15.848–50: 159

Remedia Amoris
385–6: 61

Persius
6.18–19: 84

Petronius
35.1–5: 34–5
39.5–15: 35–6

Plautus
Rudens
1–82: 65, 133
83–4: 133
358–60: 133

Pliny The Elder
2.92: 66
2.93–4: 70, 157, 160, 161–2
2.105–12: 131
18: 138–40
18.214: 138–9
18.225: 140
18.237: 139
18.269–72: 139–40
18.278–85: 131
35.199: 11, 68
36.72: 102

Plutarch
Alexander
73.1–5: 194

Caesar
69.4: 162

Marcellus
19.4–6: 65

Marius
42.4: 66

Romulus
12: 85

Sulla
37.1: 66

Propertius
1.22.9–10: 183, 185
2.24.35–8: 183
2.27.1–4: 187
2.28A.4: 141
2.34.55: 183, 185
3.18.31–2: 164
3.18.33–4: 158

4.1: 174–87
4.1A.61–6: 174–5, 183–4
4.1A.67–70: 176
4.1B.71–4: 175–6, 178, 180
4.1B.75–80: 176–7, 178
4.1B.79–86: 178–9
4.1B.86: 98
4.1B.89–102: 179–80
4.1B.89: 177
4.1B.90: 180
4.1B.92: 177
4.1B.93–6: 180
4.1B.98: 177
4.1B.99: 177
4.1B.103–8: 181, 186
4.1B.119–46: 181–5
4.1B.122: 184
4.1B.125–6: 181, 184
4.1B.147–50: 177, 185–6
4.6.59–60: 154, 168

PTOLEMY
Tetrabiblos
1.2.6–7: 173

QUINTILIAN
1.10: 111

SENECA THE ELDER
Suasoriae
4.1–3: 6, 7, 193–4
10 *praef.* 13: 193

SENECA THE YOUNGER
De Beneficiis
7.1.5: 84

Phaedra
1098–9: 169

SERVIUS
Commentary on Aeneid
4.519: 120
6.790: 71, 162
6.848: 121
8.681: 71, 157, 162

Commentary on Eclogues
9.46: 70, 71, 157, 160, 162

SEXTUS EMPIRICUS
Against the Mathematicians
5.1.2: 129

5.83–5: 192
5.88–9: 192
5.99: 192

SOLINUS
1.18: 85

STRABO
1.1.13: 111
2.5.1: 111

SUETONIUS
Augustus
5.1: 97
31: 103–4
89.2: 110
94.5: 67
94.12: 72–4, 97, 194
98.4: 105

Caligula
19.3: 194

De Grammaticis et Rhetoribus
20: 126

Iulius
56: 91
81: 70
88: 70, 162

Tiberius
14.4: 105

TACITUS
Annals
2.27–32: 105
6.20–1: 105, 194

Histories
1.22: 2, 107–8

TIBULLUS
1.1.27–8: 141
1.4.6: 141
1.4.42: 141
1.7.21: 141
2.5.71: 160

VALERIUS MAXIMUS
1.3.3: 65–6
3.2.19: 172

4.5.6: 164, 172
6.9.15: 172
8.11.2: 70
9.12.6: 68
9.15.1: 172

VARRO
Antiquitates
fr. 20: 156

VELLEIUS PATERCULUS
2.58.4: 92

VIRGIL
Aeneid
2.3: 39
3.141: 141
3.359–61: 119–20
4.517–20: 120
5.519–34: 158
6.703–51: 164
6.741–2: 166
6.847–53: 120–1
8.681: 158, 160
9.429: 120
10.175–7: 119–20
10.273–4: 141

Eclogues
9.1–4: 161
9.46–50: 160–1
10.68: 186

Georgics: 141
1.32–5: 97
1.176–7: 39
1.487–8: 160
2.353: 141
4.425–6: 141

VITRUVIUS
1 *praef.* 1: 109, 115
1 *praef.* 2–3: 110, 116, 170–1
1.1.3: 110, 112
1.1.10: 112
1.1.11: 111
1.1.12: 111
1.1.18: 109
4 *praef.* 1: 110
5 *praef.* 2–5: 109, 110
5.4.1: 115
6 *praef.* 7: 109
6.1.11: 115
6.7.7: 109
9: 112–17
9 *praef.* 18: 112
9.1–5: 112–13
9.1.1: 115
9.1.2: 113
9.1.5: 113
9.1.15: 113
9.5.4: 115
9.6.2–3: 113–15
9.7.7: 113
9.8.15: 116
10.16.1–2: 115

General Index

Aeneid (Virgil) 119–21
Alexander the Great 159, 193–4
Anaxilaus (diviner) 103
Apotheosis, Roman, dynamics of 154–8, 162–70, 170–2
Aratea (Cicero) 133–8, 141, 142
  date and structure of 133–4
Aratea (Germanicus) 68, 99, 128, 143–9, 190–1
  date and structure of 143, 146
  discretion in 148
Aratus; *see Phaenomena*
Architect / Architecture, ancient 110–12
Arellius Fuscus 6, 193–4
Ars Amatoria (Ovid) vii, 57–9, 60–1
Astrologers 4, 65–6, 69–70, 72, 74, 82–3, 104–6, 123, 127–8, 173–4, 175–87
  expulsion from Rome of 6, 65–6, 102–4, 105–6, 122–3
  professional materials of 178–9
Astrology
  after Augustus 193–7
  ancient 'handbooks' on 5, 56
  astrometeorology 1, 3–4, 127, 129–49, 189–90
  at Rome 1–2, 65–74, 97–108, 193–7
  Augustus' approach to 2, 11, 72, 102–4
  catarchic 1, 29–30, 140; *see also Astronomica*, catarchic astrology in
  definition of 1
  'hard' / strongly deterministic form of 30, 84, 86, 89
  horoscope of Rome 85, 192
  horoscopic 1, 26, 81
  philosophical criticism of 6–7, 83–6, 196–7
  (changing) relationship with Roman religion 80–3, 86, 89, 124–5
  'soft' / non-deterministic form of 88–9, 92–3
  stars as signs or causes of weather 129–49
  stars rising from / setting into the sea 48–9
  types of 1
Astrometeorology; *see* Astrology, astrometeorology
Astronomica (Manilius)
  absence of planets from 26–7, 32, 44–5, 52–3, 60, 128, 191
  and Aratus 16, 22, 27, 42–3, 54, 190–1
  and didactic poetry 12–13, 17–18, 39–40, 44, 54, 56–9
  and Firmicus Maternus 26, 37, 60
  and Germanicus 190–1
  and Lucretius 22, 29, 31, 40–3
  and Ovid 57–9, 60–1, 190–1
  Andromeda digression in 46–52
  Ascendant in 29–31
  catarchic astrology in 29–30
  contradictions / inconsistencies in 24, 25, 29–31, 50–2, 56–7
  date of 11, 14
  forward references in 27
  genre of 12
  lacuna in 52
  politics in 59–61
  readers internal to and external from 15–16, 18–19, 20–1, 26, 59
  stars imparting traits to natives in 33–6, 45–6
  Stoicism in 21–4, 27, 42, 49, 52
  (deteriorating) teacher / student relationship in 12–13, 15, 16, 19–21, 25–8, 31–3, 33–44, 44–52, 53–5
Athenion 66
Augustus / Octavian 2–4, 11, 71–4, 97–108, 110, 124–5, 144–5, 148
  and Apollo 99, 102
  and Capricorn 97–100, 123–4, 145
  and Libra 97, 123–4
  horoscope of 72–4, 103, 104
  Res Gestae 106–7

Balbillus (astrologer) 106
Barton, Tamsyn 5–6
Belvedere Altar 168

Caesar (G. Iulius Caesar) 67, 69–70, 70–1, 74, 77, 85, 86, 88, 92–3; *see also* Caesar's comet; *De Astris*
  catasterism of 3, 4, 151–72
  divinity won through earthly achievements and / or divine agency 153–4, 170–2
  praised for superiority of son (Augustus) 153–4
  stellar imagery of 157, 159
Caesar's comet 4, 70–1, 74, 89, 91–3, 98, 151–72, 195
  as a sign of or physical manifestation of Caesar 161–2
  as a star and / or a comet 158–9
Calpurnius Siculus 195
Capricorn; *see* Augustus / Octavian, and Capricorn
Carneades (of Cyrene) 68
Cicero (M. Tullius Cicero) 3, 68, 69, 75–93, 111, 156–7, 162–3, 189; *see also De Consulatu Suo*; *De Divinatione*; *De Fato*; *De Re Publica*; *Philippics*
Comets; *see also* Caesar's comet; Halley's comet
  etymology of 167–8
  predictions of violent change from 70–1
Cosmos and imperium, motif 115
Cramer, Frederick 5
Crassus (M. Licinius Crassus) 67, 85

De Architectura (Vitruvius) 3, 109–17, 149, 170–1, 189
  date of 109
  purpose of 110–12
De Astris (Caesar) 138–40
De Astronomia (Hyginus) 3, 125–7
  date and nature of 125–6
De Consulatu Suo (Cicero) 81, 134
De Divinatione (Cicero) 75–86, 191–2, 196
  date and structure of 75–6
  overlap between Cicero and 'Marcus' in 76–80
De Fato (Cicero) 86–93
  date of 86–7
De Re Publica (Cicero) 156–7, 162–3
Didactic poetry; *see Astronomica*, and didactic poetry
Dog Star 130–3, 135, 139–40, 141–2, 147, 148, 191

Ennius (Q. Ennius) 154–6, 162
Eratosthenes (*Catasterismoi*) 126, 149
Euhemerus (of Messene) 154–6

Fasti (Ovid) 127–8, 168–9, 171, 190–1
Favorinus (philosopher) 93, 195–7
Firmicus Maternus (*Mathesis*) 173–4; *see also Astronomica*, and Firmicus Maternus

Gaius Gracchus (Gaius Sempronius Gracchus) 68
Gellius (Aulus Gellius) 184, 195–7
Geminos (*Introduction to the Phenomena*) 131–2
Gemma Augustea 99
Genethliology *see* Astrology, horoscopic
Germanicus (Germanicus Iulius Caesar); *see Aratea* (Germanicus)

Halley's comet 66
Haruspicy 67–8, 70–1, 74, 83
Herbert-Brown, Geraldine 6–7
Herrenius Siculus (haruspex) 68
Hipsalus (Cn. Cornelius Hipsalus) 65–6
Hirtius (Aulus Hirtius) 86–7, 90–1
Horace (Q. Horatius Flaccus) 3, 121–5, 142–3; *see also Odes*
Horologium Augusti 99, 101–2, 115
Horos 4, 174–87
Hyginus (C. Iulius Hyginus); *see De Astronomia*

Laberius (Decimus Laberius) 67
Libo Drusus (M. Scribonius Libo Drusus) 105
Libra; *see* Augustus, and Libra
Lucretius (T. Lucretius Carus); *see Astronomica*, and Lucretius
Lygdamus 141

Manilius (Manilius Antiochus) 11, 68
Manilius (Marcus Manilius) 2–3, 11–61; *see also Astronomica*
  life of 11
Marcellus (M. Claudius Marcellus) 65
Metamorphoses (Ovid) 151–4, 155, 162–8, 169–70, 171–2

Nigidius Figulus (P. Nigidius Figulus) 67, 81, 84, 98, 126

Octavian (C. Octavius); *see* Augustus / Octavian
Octavius (Gn. Octavius) 66, 68
Odes (Horace) 121–5
  date of 122
Ovid (P. Ovidius Naso) 3, 68, 99, 141–2; *see also Ars Amatoria*; *Astronomica*, and Ovid; *Fasti*; *Metamorphoses*
Oxyrhynchus 178–9

Panaetius (of Lindus) 68
Pantheon 100–1
Parapegma 129, 138
Perusia 185
Phaenomena (Aratus) 16, 22, 27, 42–3, 54, 68, 113, 116, 117, 124, 126, 130, 133–8, 141, 143–6, 149, 189, 190–1
Philippics (Cicero) 87, 92
Planets, absence from Augustan literature of 128, 191; *see also Astronomica*, absence of planets from
Pompey (Gn. Pompeius Magnus) 67, 68, 85
Posidonius (of Rhodes) 68
Postumius (C. Postumius) 68
Propertius 4, 141, 173–87, 190, 196
Ptolemy (*Tetrabiblos*) 148, 173

Quintus, fragment of *Aratea* attributed to 137

Res Gestae; *see* Augustus / Octavian, *Res Gestae*

Seneca the Elder 193–4
Sidus Iulium; *see* Caesar's comet
Sirius; *see* Dog Star
Soul, distinct from mortal body 162–3, 164
Soul, similar / different in shape to / from body 164, 165, 168
Spurrina (haruspex) 70
'Student experience' viii
Suasoriae (Seneca the Elder) 193–4
Sulla 66, 68

Tarutius Firmanus (L. Tarutius Firmanus) 85, 192
Theogenes (astrologer) 72
Thrasyllus (astrologer) 104–6, 194
Tiberius 11, 14, 104–6, 143, 172
Tibullus (Albius Tibullus) 141
Trimalchio 34–6

Umbria 185

Valerius Maximus 172
Varro (M. Terentius Varro) 69, 77, 85
Ventidius Bassus (P. Ventidius Bassus) 70
Vettius Valens 148, 183
Virgil (P. Vergilius Maro) 3, 141, 160–1; *see also Aeneid*
Vitruvius (M. Vitruvius Pollio) 3, 149; *see also De Architectura*
  life of 109
  status as architect 111–12, 117
Vulcanius (haruspex) 70

The manufacturer's authorised representative in the EU for product safety is Oxford University Press España S.A. of El Parque Empresarial San Fernando de Henares, Avenida de Castilla, 2 - 28830 Madrid (www.oup.es/en or product.safety@oup.com). OUP España S.A. also acts as importer into Spain of products made by the manufacturer.
Printed and bound by CPI Group (UK) Ltd, Croydon, CR0 4YY
14/09/2026
02216304-0001